W9-CRB-258

Hans Christian Oersted

Friedrich Wöhler

Henri Ste. Claire Deville

Paul Louis T. Hèroult

Charles Martin Hall

Pioneers of Aluminium Production

# ALUMINIUM AND ITS ALLOYS

# ELLIS HORWOOD SERIES IN METALS AND ASSOCIATED MATERIALS

*Series Editor:* E. G. WEST, OBE, Independent Metallurgical Consultant

An authoritative series of books on metallurgy and materials science for teaching, research and industry, serving engineers in production and design, management scientists and research workers.

**TIN AND ITS ALLOYS AND COMPOUNDS**
B. T. K. BARRY and C. J. THWAITES, International Tin Research Institute, UK
**COBALT AND ITS ALLOYS**
W. BETTERIDGE, former Chief Scientist, International Nickel Limited, UK
**NICKEL AND ITS ALLOYS**
W. BETTERIDGE, former Chief Scientist, International Nickel Limited, UK
**IRON CONTROL IN HYDROMETALLURGY**
J. E. DUTRIZAC, Head of Metallurgical Chemistry Section, CANMET, Ottawa, and A. J. MONHEMIUS, Royal School of Mines, Imperial College of Science and Technology, London
**PRINCIPLES OF HYDROMETALLURGICAL EXTRACTION AND RECLAMATION**
E. JACKSON, Principal Lecturer in Metallurgy, Sheffield City Polytechnic, Sheffield
**ALUMINIUM AND ITS ALLOYS**
F. KING, former Technical Manager, British Alcan Testing Laboratories, Royerstone
**ZINC AND ITS ALLOYS AND COMPOUNDS**
S. W. K. MORGAN, former Managing Director, Imperial Smelting Processes Ltd.
**COPPER AND ITS ALLOYS**
E. G. WEST, OBE, former Director, Copper Development Association, London
**BASIC CORROSION AND OXIDATION, Second Revised Edition**
J. M. WEST, Department of Metallurgy, University of Sheffield

# ALUMINIUM AND ITS ALLOYS

FRANK KING, M.I.M.
former Technical Manager, British Alcan Testing Laboratories

**ELLIS HORWOOD LIMITED**
Publishers · Chichester

Halsted Press: a division of
**JOHN WILEY & SONS**
New York · Chichester · Brisbane · Toronto

*0 2880672*

First published in 1987 by
**ELLIS HORWOOD LIMITED**
Market Cross House, Cooper Street,
Chichester, West Sussex, PO19 1EB, England
*The publisher's colophon is reproduced from James Gillison's drawing of the ancient Market Cross, Chichester.*


**Distributors:**

*Australia and New Zealand:*
JACARANDA WILEY LIMITED
GPO Box 859, Brisbane, Queensland 4001, Australia

*Canada:*
JOHN WILEY & SONS CANADA LIMITED
22 Worcester Road, Rexdale, Ontario, Canada

*Europe and Africa:*
JOHN WILEY & SONS LIMITED
Baffins Lane, Chichester, West Sussex, England

*North and South America and the rest of the world:*
Halsted Press: a division of
JOHN WILEY & SONS
605 Third Avenue, New York, NY 10158, USA


© 1987 F. King/Ellis Horwood Limited

**British Library Cataloguing in Publication Data**
King, Frank, *1913*–
Aluminium and its alloys. —
(Ellis Horwood series in metals and associated materials).
1. Aluminium
I. Title
669'.722    TN775
**Library of Congress CIP data also available**

ISBN 0–7458–0013–0 (Ellis Horwood Limited)
ISBN 0–470–20849–X (Halsted Press)

Phototypeset in Times by Ellis Horwood Limited
Printed in Great Britain by The Camelot Press, Southampton

**COPYRIGHT NOTICE**
All Rights Reserved. No part of this publication may be reproduced, stored in a retrieval system, or transmitted, in any form or by any means, electronic, mechanical, photo-copying, recording or otherwise, without the permission of Ellis Horwood Limited, Market Cross House, Cooper Street, Chichester, West Sussex, England.

TN775
K461
1987
CHÉM

# Table of contents

**Chapter 7 — Manufacturing Processes**

# Author's preface

Aluminium has grown in little more than a century from virtually a chemical curiosity to the world's second most commonly used metal. Its existence was postulated by Davy in the earliest years of the nineteenth century; its separation was first achieved some twenty years later; its extraction by reduction with molten sodium was undertaken commercially from 1855 to 1890 and the modern electrolytic process dates only from 1886. Since that time, mining of aluminium ore, bauxite, the extraction from it of alumina by chemical processing, electrolytic smelting of the metal and its processing into fabricated forms have spread throughout the world. There is a wide and varied family of aluminium alloys now used for a multitude of purposes from the thinnest foil for wrapping through every engineering industry to high technological applications in aeronautics, space exploration and electronics.

The aluminium industry has its foundations in basic research and development activities which continue on a wide scale both to assist production of the metal and its alloys and in its continuing application in virtually all the major industries.

The aim of this book is to provide an authoritative account of aluminium, its present technological position and its industrial capabilities, as well as a quick and ready source of reference data. It is intended for engineers concerned with production and design, for management in planning to extend the range of materials and products, and for students and teachers of engineering as well as in metallurgy and materials science. The occurrence and processes for extraction of aluminium are described with the economic–technological background involved and this is followed by summaries of the methods of fabrication applied to the aluminium group of metals.

The aluminium alloys are discussed in terms of their equilibrium diagrams and heat treatment within a range from the metal itself with its low strength through the range of alloys to those eight to ten times stronger than the pure metal in the soft condition.

It is expected that through the references and recommendations for further reading, users and students will be encouraged to follow up this introduction to the possibilities for present and future applications of the light alloys of aluminium. It is a further addition to 'The Industrial Metals' series and follows the same general pattern.

*Note*: SI units are used throughout with conversion tables to Imperial Units in Appendix 5.

## ACKNOWLEDGEMENTS

Thanks are gladly given to those who have contributed illustrations or have given permission for diagrams and tabulated material to be used from their works, in particular the Aluminium Federation (formerly the Aluminium Development Association) for Figs 3, 23, 54, 57 and 59 reproduced from their publication *The Properties of Aluminium and its Alloys* and for much material culled from tables in the same book. The equilibrium diagrams in Chapter 4 are based mainly on the ADA Information Bulletin No. 25. Other illustrations and data have been provided by British Alcan, Alcan International Ltd, Aluminium Standards Association and the British Standards Institution. Thanks are also given to the following individuals for their help in a similar manner: G. F. Hancock, M. Van Lancker, E. I. Brimelow, H. H. Read, A. Von Zeerleder, M. Duraut, J. F. Hawkins, G. Camatini, A. Prince and to the editor of the *Series in Industrial Metals*, Dr E. G. West, and his daughter, Philippa, for assistance in producing this book.

F.K.

# 1

# Introduction and basic properties of aluminium

The earliest successful methods for the production of metallic aluminium involved the use of potassium amalgam or metallic sodium, as will be seen from Table 1. The cost of production was high and this, combined with the low mechanical strength of pure aluminium, limited the uses to which the metal could be applied. However, the invention of the electrolytic process for the reduction of oxide of aluminium, alumina, opened up the possibility for reducing costs.

It is about one hundred years since Paul Louis Hèroult in France and Charles Martin Hall in America described, in their basic patents of 1886, the electrolytic process which is still in use today. These two inventors had never met nor corresponded but the commercial development of their patents in Europe and the USA led, of course, to law suits. To complete the coincidences, they were both the same age (22) in 1886 and both died in 1914.

In early experimental work on the electrolytic reduction of salts of aluminium, Sir Humphrey Davy in 1807 used a galvanic battery of one thousand cells but without success. The availability of cheaper electric power in the years following the installation of the first power station in 1881 contributed to the industrial exploitation of the Hall–Hèroult process.

It was the use of aluminium for such applications as castings for domestic utensils, rolled and drawn wire for electrical power distribution and sections for engineering purposes together with rolled sheet products which provided outlets for the increasing amounts of metal which became available.

To improve the mechanical properties of aluminium sheet products, aluminium–copper was one of the first alloys used at the beginning of the present century but the presence of copper in the alloy reduced the very good corrosion resistance of the metal. Many other strain hardening binary and ternary alloys of aluminium with such metals as zinc, magnesium or silicon, together with the iron and silicon present as impurities from the

**Table 1** — Summary of discovery and early development of aluminium reduction processes

| Year | Investigator and country | Process | Product | Price per kilogram |
|---|---|---|---|---|
| 1807 | Sir Humphrey Davy (Great Britain) | Electrolysis of fused mixture potash and alumina | Unsuccessful | — |
| 1825 | Hans Christian Oersted (Denmark) | Heating together aluminium chloride and potassium amalgam | Aluminium amalgam. Small lump of white metal | — |
| 1825 | Fogh (France) | A repeat of the above procedure | Aluminium in compact crystal form | — |
| 1827 | Friedrich Wöhler (Germany) | Heating metallic potassium and aluminium chloride | A few grains of metal as grey powder | — |
| 1845 | Friedrich Wöhler (Germany) | Reduction of aluminium chloride with potassium vapour | Metallic globules of aluminium | |
| 1854 | Henri Sainte Claire Deville (France) | First commercial process. Reduction of double chloride of sodium and aluminium with metallic sodium, prepared separately by reduction of sodium carbonate with charcoal | 97–97.3% Aluminium | |
| 1856 | A. Monnier (America) | Reduction by sodium in New Jersey | Aluminium | — |
| 1857 | Deville and P. Morin (France) | Full-scale plant at Nanterre using sodium reduction | Aluminium | c. £44.00 |
| 1858 | W. Gerhard (Britain) | Plant in London for reduction by sodium | Aluminium | £40.00 |
| 1858 | Tissier Bros (France) | Plant at Rouen for sodium reduction of aluminium chloride | Aluminium | |
| 1860 | H. Merle & Co. (France) | Plant at Salindres for reduction by sodium | Aluminium | |

| Year | Name/Company | Description | Product | Price |
|---|---|---|---|---|
| 1860 | Bell Bros (Britain) | Plant at Co. Durham for reduction by sodium | Aluminium | |
| 1863 | Wirz & Co. (Germany) | First aluminium reduction plant in Germany | Aluminium | |
| 1867/77 | J. F. Webster (Britain) | Plant near Birmingham for reduction by sodium followed in 1886/9 by larger plant using cheap sodium from Castner process | 99% Aluminium | |
| 1884 | Col. Frismuth (USA) | Plant at Philadephia using sodium reduction process | Aluminium | $15–$30 |
| 1886 | Paul Louis Toussaint Héroult (France) / Charles Martin Hall (America) | Electrolysis of a fused mixture of alumina in cryolite | Aluminium | £0.44 |
| 1887 | Héroult | First operated at Neuhausen on Rhine by predecessor of the Swiss Company (now Aluminium Industrie A.G. — AIAG) | Aluminium alloys | |
| 1888 | Hall | Pittsburgh Reduction Company using Hall's process in USA. In 1907 the Company changed its name to the Aluminum Company of America — ALCOA | Aluminium | $2–1.3 |
| 1888 | Webster | Cheap sodium produced by the Castner process used by Webster | Aluminium | £3.50–£2.0 |
| 1889 | Héroult | AIAG plant established | Aluminium | |
| 1896 | Héroult process | First reduction plant in Britain using hydroelectric power in Scotland (British Aluminium Co.) | Aluminium | £1–£0.5 |

reduction process, were offered to widen the scope of possible applications for the metal. The claims made for some of these alloys were so extravagant that a contemporary declared that they tended to handicap rather than promote the exploitation of the metal.

It was in 1907 that Wilm discovered that certain alloys of aluminium with copper and magnesium responded to thermal treatment at elevated temperatures of the order of 500 °C, followed by quenching in cold water and ageing at ordinary temperature for a period of days to give a significant increase in hardness. This led to the use of the metal for building the structures of airships and parts for aeroplanes. It was several years later, in the early 1920s, that Dr Marie Gayler and her colleagues at the National Physical Laboratory in England, as well as Paul Merica *et al.* in the USA, offered an explanation for the phenomena.

Since Wilm's discovery a large range of alloys has been developed which respond to heat treatment and quenching followed by ageing at room temperature or at temperatures in the range 100–200 °C or more, depending upon composition, to give enhanced combinations of mechanical strength and other properties such as improved corrosion resistance or notch toughness.

Aluminium alloys with 0.5 per cent magnesium and 0.5 per cent silicon can be extruded to very intricate thin sections having medium strength. This type of alloy is used extensively for architectural applications. The aluminium–copper–magnesium–manganese–silicon alloys and those of aluminium–zinc–magnesium–copper type respond to solution heat treatment followed by ageing at normal temperature or temperatures in the range 120–200°C to give good mechanical properties and are used for aircraft construction in the form of sheet, plate, extrusions, rod and wire.

### Commercial development

During the periods of the two World Wars there were big increases in the demand for aluminium and these were followed by slumps in sales. Reduction capacity having already been installed to meet war-time requirements, there was every incentive to find outlets for the production capability of these plants. After the Second World War this challenge was met not so much by the development of new alloys as by the search for new uses for existing alloys and by the development of practices and processes designed to improve quality and productivity. Large tonnages of metal reclaimed from wrecked and obsolete aircraft were used for the fabrication of factory built temporary housing and at the same time there was increasing demand for sheet for the manufacture of domestic utensils and alloy products for road vehicles, structural engineering and buildings, electrical power distribution and marine applications.

Over the comparatively short history of the commercial use of aluminium there has been a proliferation of alloy compositions introduced to meet the specific requirements of a small section of the market or of industrial customers or to use readily available materials such as reclaimed metal. To further confuse matters, individual firms had their own alloy identification systems and countries have their own national specifications and standards. These practices have given rise to complications when attempting to assess the suitability of material from one source with that from another, or of determining the applicability of the information published on a given alloy to material having a different designation. In recent years there has been concerted international action to adopt a unified system of alloy designations and for individual countries to draft their national specifications in such a way that material in compliance with national requirements will also meet the internationally agreed limits for material having the same alloy designation, thus facilitating the import and export of products. The subject of specifications is dealt with in more detail at the end of this chapter.

## 1.1  PROPERTIES OF PURE ALUMINIUM

### 1.1.1  Physical properties
The atomic and nuclear properties are given in Table 2 and its commoner

**Table 2** — Atomic and nuclear properties [1]

| Property | Value | Source |
|---|---|---|
| Atomic number | 13 | |
| Atomic weight | 26.9901 | |
| Atomic radius at 25°C | 1.42885 Å | |
| Isotopes | None | |
| Valency | 3 | |
| Ionic radius $Al^{+3}$ | 0.57 Å | Goldschmidt |
| | 0.50 Å | Pauling |
| Ionisation potential | | |
| Work function (eV) | 42 | |
| Neutron absorption cross-section | | |
| Reaction δab | 230 ± 5 | |
| Scattering δ | 1.4 ± 0.1 | |

physical properties in Table 3.

   In the Periodic Table, aluminium, with its atomic number 13, is in Group III, the same as boron, gallium and indium. It falls in the second short period after sodium and magnesium and is followed by silicon.

**Table 3** — Physical properties of pure aluminium

| Property | Value | |
|---|---|---|
| Colour — reflected light | Silvery white | |
| Crystallographic structure | Face-centred cube | |
| Lattice constant $\alpha$ at 25°C | 0.40414 nm | |
| Minimum inter-atomic distance | 0.28577 nm | |
| Slip plane — primary twinning plane | 1, 1, 1 | |
| Density at 20°C | 2.699 g/cm$^3$ | |
| Solid at 600°C | 2.55 g/cm$^3$ | |
| Liquid just above melting point 600°C | 2.38 g/cm$^3$ | |
| Volume change on solidification | 6.7% | |
| Casting contraction (linear) | 1.7–1.8% | |
| Heat of combustion | 200 Kcal per g-atom | |
| Melting point | 660.2°C | |
| Boiling point | 2057 °C/2480°C | |
| Vapour pressure at 1200°C | $1 \times 10^{-3}$ mm mercury | |
| Specific heat | J(kgK) | cal/g/°C |
| at 20°C | 930 | 0.222 |
| at 100°C | 935 | 0.223 |
| at 500°C | 1110 | 0.266 |
| Mean specific heat 0 to 658°C solid | 1045 | 0.25 |
| | J/g | cal/g |
| Heat of fusion | 387 | 92.4 |
| Heat of vaporisation | 8200–8370 | 7950–2000 |
| Linear coefficient of thermal expansion $\times 10^6$ | 23.0 | (20°–100°C) |
| Linear coefficient of thermal expansion $\times 10^6$ | 28.1 | (20°–600°C) |
| Linear coefficient of thermal expansion $\times 10^6$ | 31.1 | (at 500°C) |
| Viscosity at melting point | 130 m Nsm$^{-2}$ | |
| Surface tension at melting point | 914 m Nm$^{-1}$ | |
| Thermal conductivity | W/mK | cal/s/cm$^2$/cm/°C |
| at 0°C | 209 | 0.50 |
| at 100°C | 213 | 0.51 |
| at 200°C | 217 | 0.52 |
| Electrical conductivity at 20°C International Annealed Copper Standard (IACS) = 100 | 65.5% | |
| Electrical resistivity at 20°C | 2.69 $\mu\Omega$cm | 2.63 ohm cm |
| Temperature coefficient of electrical resistance at 20°C | 0.0041 | |
| Thermo-electric force (relative to Pt with cold junction at 0°C) | +0.416 at 100°C | |
| Magnetic susceptibility at 18°C $\times 10^6$ | 0.63 | |
| Standard potential at 25°C | −1.69 V | |
| Electrochemical equivalent Al$^{+++}$ | 0.3354 g/A/h | |
| Velocity of sound in aluminium | 5100 m/s | |

Aluminium consists of a single isotope of mass number 27 (abundance 100 per cent) and isotopic number (i.e. neutrons–protons) 1. Its nucleus contains 13 protons and 14 neutrons, with its 13 electrons distributed as follows:

$$(\underline{1}s)^2; \quad (\underline{2}s)^2; \quad (\underline{2}p)^6; \quad (\underline{3}s)^2; \quad (\underline{3}p)^1$$

The index gives the number of electrons, the underlined number the principal electron number ($n$) and the letter the secondary electron number s or p. The outer shell contains three electrons, two of which are of the $(3s)^2$ type; this explains the existence of monovalent aluminium. The 'free' atom has only one valency electron; but if atoms are bonded together in the solid lattice or in the molten state, hybridisation takes place and the three valency electrons can all behave identically [1].

### 1.1.1.1  Colour
Aluminium is a silvery-white-coloured metal having high reflectivity for light and heat. The alloys of aluminium are generally of a similar colour, some with a bluish tinge.

### 1.1.1.2  Density
The density of aluminium is 2.699 g/cm$^3$ against the calculated value of 2.694. It falls to 2.55 g/cm$^3$ for the solid at 660°C, just below the melting point, and 2.38 g/cm$^3$ for molten metal at this temperature. Fusion is accompanied by an increase in volume of 6.5–6.7 per cent depending upon the metal purity, the lower value corresponding to 99.5 per cent aluminium. (The comparable volume change for copper is 4 per cent.)

The addition of silicon or magnesium to form alloys results in reduction of density to 2.65 for 12 per cent silicon alloy and to 2.55 for a 10 per cent cast magnesium content or 2.63 for 7 per cent wrought aluminium–magnesium alloy.

### 1.1.1.3  Thermal properties
The melting point of 99.99 per cent aluminium is 660.2°C and the heat of fusion is 10,800 J (2480 calories) per gram-atom (26.99 g) or 387 J/g.

The thermal conductivity of aluminium at 209 W/mK is about 52.5 per cent that of pure copper which is 399 W/mK.

### 1.1.1.4  Electrical properties
The electrical conductivity of aluminium is normally expressed as a percentage of the International Annealed Copper Standard (IACS) adopted in 1913. The copper standard of 100 per cent conductivity is equivalent to an electrical resistivity of 0.017241 $\Omega$mm$^2$/m (1.7241 microhm/cm cube). Pure aluminium has a resistivity of 2.630 $\Omega$mm$^2$ which is equivalent to a conductivity of 63.8 per cent IACS. Small amounts of impurities and in particular titanium, vanadium and chromium have a deleterious effect upon conductivity, but when converted into diborides of the type $TiB_2$, $VB_2$ and $CrB_2$ by treatment of the molten metal with boron salts or Al–B master alloy, these impurities are not harmful from an electrical conductivity aspect.

For general electrical applications, aluminium normally has iron and silicon present in amounts sufficient to improve tensile strength whilst conductivity is maintained at between 60 and 62.5 per cent IACS. Certain alloys containing small amounts of magnesium and silicon give conductivity values after appropriate heat treatment of 53 to 56 per cent IACS, with enhanced mechanical properties. Fig. 1 shows some typical values.

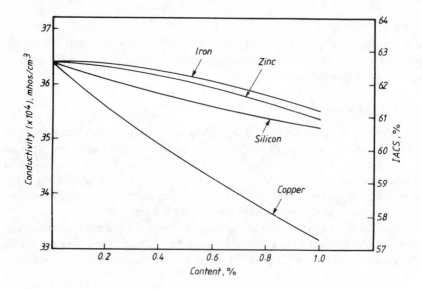

Fig. 1 — The effect of conductivity of aluminium (99.93% purity), of iron, silicon, copper and zinc. (*After Von Zeerleder.*)

### 1.1.1.5 Optical properties

A clean surface of pure aluminium reflects from 80 to 85 per cent incident visible radiation. Surface treatment by means of special electrical and chemical techniques is required to develop the full reflecting power of aluminium.

The reflective power of aluminium is of importance in the construction of various types of light or heat reflectors. It also means that aluminium vessels will take up less heat in sunshine than those of other metals such as copper and steel.

By appropriate oxidation it is possible to give aluminium one of the best radiating surfaces known, a characteristic of importance in heat exchangers.

Highly polished aluminium has an absorptivity for solar radiation of 0.10

to 0.40. Emissivity (10 to 38°C) varies according to the wavelength of the light as follows:

| Wavelength ($\mu$m) | 10.0 | 5.0 | 1.0 |
|---|---|---|---|
| Spectral emissivity | 0.02–0.04 | 0.03–0.08 | 0.08–0.27 |

A typical value for the oxidised metal is 0.30 for light of wavelength 0.65 $\mu$m.

### 1.1.1.6  Magnetic properties
Aluminium and its alloys are slightly paramagnetic.

### 1.1.1.7  Elasticity
The modulus of elasticity of aluminium is relatively low and is sensitive to small changes in the quantity of impurities present. The value of Young's modulus for 99.997 per cent pure aluminium is 64,200 N/mm$^2$ (4150 tonf/in$^2$) and that for 99.950 per cent pure aluminium is 69,000 N/mm$^2$ (4450 tonf/in$^2$). The values for commercial grades of aluminium are slightly higher and the range of alloys is 65,000 to 80,000 N/mm$^2$ (4230 to 5270 tonf/in$^2$) (see Fig. 2).

The elastic extension of aluminium and its alloys, under a given stress, is thus about three times as great as that for steel. This is advantageous in structural members required to withstand impact. The torsion modulus is 25,000 n/mm$^2$ and Poisson's ratio is 0.34. Table 4 summarises the elastic properties.

**Table 4** — Elastic properties of aluminium [3]

| | | |
|---|---|---|
| Modulus of elasticity | | |
| 99.99% pure Al | 64,200 N/mm$^2$ | ($8.45 \times 10^6$ lbf/in$^2$) |
| 99.950% pure Al | 69,000 N/mm$^2$ | ($9.07 \times 10^6$ lbf/in$^2$) |
| Modulus of rigidity | 17,000 N/mm$^2$ | ($3.85 \times 10^6$ lbf/in$^2$) |
| Poisson's ratio | 0.32–0.36 | — |

### 1.1.1.8  Diffusion
The self-diffusion of aluminium depends on crystal structure and crystal size: it may also be affected by surface effects. The self-diffusion coefficients at two ranges of temperatures are:

| Temperature (°C) | A | Q |
|---|---|---|
| 300–650 | 2.23 | 34.5 |
| 450–650 | 1.71 | 34.0 |

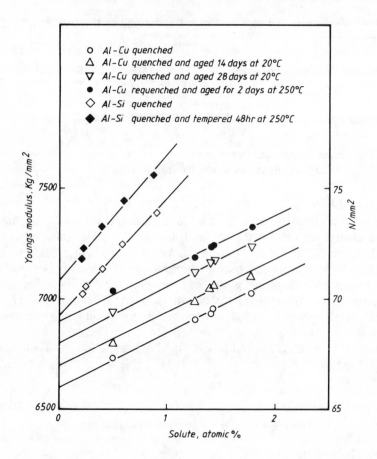

Fig. 2 — Variation of Young's modulus with solute content in w% composition for quenched and for aged solid solutions of copper or silicon in aluminium. (*After Van Lancker.*)

There is information available on the diffusion of a number of elements in aluminium [2].

### 1.1.2 Chemical properties

Corresponding to its position in the electrochemical series of elements in Table 5 and to the heat of formation of its oxide — $10^6 \times 1.606$ J/gmol (383,900 cal per g-mol) — aluminium is classed as a very reactive and easily oxidisable element. However, in practice it proves to be very resistant to corrosion, both in the form of pure metal and when alloyed with the metals

**Table 5** — Electrochemical relationships between metals and alloys [3]

| Electromotive series[†] | Galvanic series[‡] |
|---|---|
| Magnesium | Magnesium |
| Aluminium | Zinc |
| Zinc | Aluminium and Al–Mg–Si alloys |
| Chromium | Cadmium |
| Iron | Al–Cu–Mg–Si alloys (H15 type) |
| Cadmium | Iron and mild steel |
| Nickel | Lead |
| Tin | Tin |
| Lead | Brass |
| Copper | Nickel |
| Silver | Bronze |
| Mercury | Copper |

[†] Based on the molar electrode potential of a metal against its own salt solution.
[‡] Based on electrode potentials in sea water and salt solutions.

iron, manganese and magnesium. The chemical resistance is due to the formation of a very thin, compact and firmly adherent oxide film which is insoluble in water and many other chemcials. It is this oxide film which protects the underlying metal from further attack even though the oxide layer, as formed in air, is only some $4$–$5 \times 10^{-6}$ mm thick. When heated in air for lengthy periods its thickness may increase to $2 \times 10^{-4}$ mm. By anodic oxidation (anodising) as described in Chapter 7 the oxide film may be increased to as much as 0.030 mm thickness. As formed in air or by anodising, the film ($\alpha$) is amorphous but on heating it changes to a harder crystalline form known as $\gamma$ oxide. The $\alpha$ form is able to absorb moisture from the air and the thicker anodic films are capable of absorbing and retaining dyes.

General corrosion, on which the whole metal surface is evenly attacked, occurs only under the influence of oxide solvent media which after dissolving the passive surface coating attack the metal with the formation of a salt. The strongest solvents of aluminium are the halogen acids (hydrogen chloride and hydrogen fluoride), concentrated sulphuric acid and aqueous solutions of alkali hydroxides and carbonates of sodium and potassium. The attack by alkali solutions can be inhibited by certain colloids such as sodium silicates of $SiO_2$ to $Na_2O$ ratios of at least 1 to 1. Thus, the oxide $Al_2O_3$ is amphoteric and there is a considerable body of data on the action of a large range of substances on the metal and its oxide, as given in Appendix 1.

There are many reagents which have no action on, or react only slightly with, aluminium; these include concentrated solutions of nitric acid, ammonia and most organic acids and sulphides — exceptions are formic acid and oxalic acid which are solvents of both aluminium oxide and the metal.

The chemical resistance depends upon the particular chemical and the concentration of the solution as well as upon the purity of the metal. In nitric acid, 99.99 per cent aluminium is much more resistant to attack than 99.5 per cent aluminium, whereas in dilute sulphuric acid the solubilities of the two grades of metal are very similar.

Alloying additions of up to 1 per cent of either silicon or zinc have very little effect on the corrosion resistance of aluminium but heavy metal alloying constituents, particularly copper or nickel in amounts as little as 0.1 per cent, effect a very strong increase in susceptibility to corrosion and to a lesser extent so too does iron.

The effects of additions of magnesium are particularly interesting as the composition of the oxide film varies with the magnesium content not only on the average but also according to the thickness of the film and its mode of formation. The alloys with 2–7 per cent magnesium are particularly resistant to sea water corrosion, but as discussed later (Chapter 5.6.5) above about 3 per cent magnesium, they may suffer intercrystalline attack if the $Al_3Mg_2$ ($\beta$) phase is located at the grain boundaries as a result of heat treatment.

### 1.1.2.1 Hydrogen

Aluminium does not combine with hydrogen but it is dissolved in both the solid and the liquid metal. The effect of temperature on the absorption of hydrogen by aluminium is given in Table 6 and Fig. 3. It will be seen that the

**Table 6** — The solubility of hydrogen in aluminium under an external pressure of 1 atmosphere hydrogen

| Metal temperature °C | | Solubility of hydrogen ml/100 g Al | at% |
|---|---|---|---|
| | 300 | 0.001 | $2.2 \times 10^{-7}$ |
| | 400 | 0.005 | $1.1 \times 10^{-6}$ |
| | 500 | 0.0125 | $2.8 \times 10^{-6}$ |
| Solid | 660 | 0.036 | $8.1 \times 10^{-6}$ |
| Molten | 660 | 0.69 | $1.55 \times 10^{-4}$ |
| | 700 | 0.92 | $2.07 \times 10^{-4}$ |
| | 750 | 1.23 | $2.77 \times 10^{-4}$ |
| | 800 | 1.67 | $3.74 \times 10^{-4}$ |
| | 900 | 2.80 | $6.1 \times 10^{-4}$ |
| | 1000 | 4.8 | $8.95 \times 10^{-4}$ |
| | 1100 | 5.2 | $11.6 \times 10^{-4}$ |

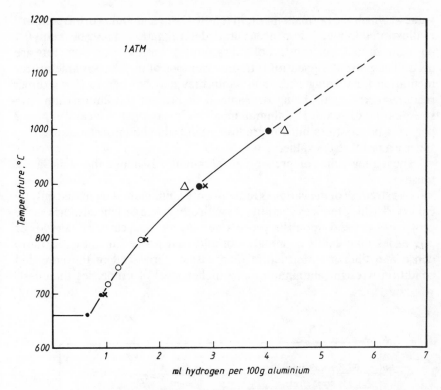

Fig. 3 — Variation in hydrogen content with temperature.

amount dissolved increases rapidly with increasing the temperature of the molten metal. To obtain sound castings, free from gas porosity, overheating of the molten metal must be avoided and special degassing treatments must be employed.

### 1.1.2.2  *Nitrogen*
Aluminium can combine with nitrogen to give aluminium nitride (AlN) which is a hard solid. The formation of this compound can be caused by overheating during melting. Its presence in castings or wrought products is detrimental to properties. Filtration of molten metal is used to remove inclusions of this type.

### 1.1.3  Mechanical properties
### 1.1.3.1  *Proof stress and tensile strength*
Like the other non-ferrous metals, aluminium and its alloys do not exhibit a sharply defined elastic limit, yield point or limit of proportionality. For quality control testing and for design purposes it is general practice to use the

value at which the permanent set, in tensile testing, amounts to 0.2 per cent, as illustrated in Fig. 4. For the accurate determination of 0.02 per cent, 0.1 per cent and 0.2 per cent proof stress values, sensitive extensometers are used. These are designed for different methods of use. By suitable instrumentation for reading stress and strain they may be used to give a direct reading of the value being determined or present the information as a stress–strain diagram. For aluminium alloys it is normal to measure the 0.2 per cent proof stress but for grades of pure aluminium it is not usual to specify a proof stress value.

The strength in compression of aluminium is about the same as in tension.

The strength of the various grades of pure aluminium is increased only by cold work which reduces ductility. As with other pure cast metals, the tensile properties depend upon the process employed; sand castings have lower properties than chill cast metal. Hot and cold worked metal has a more dense and finer grain structure than cast material and in the annealed condition wrought aluminium has a higher level of properties than cast.

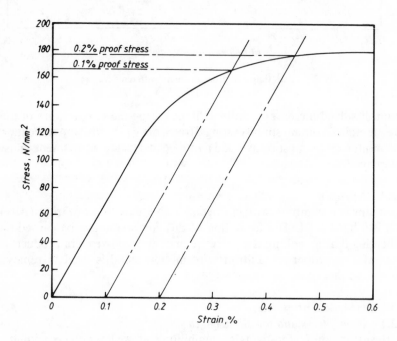

Fig. 4 — Load–extension curve, illustrating method of deriving proof stress.
(*Courtesy of Aluminium Federation.*)

Because of the temperature differences generated during the cold rolling process, temper rolled material given a reduction in few passes on a strip rolling mill has slightly lower tensile properties than that given the same overall reduction in a large number of passes through a flat sheet mill.

The shear stress of 99.8 per cent aluminium is 58 N/mm$^2$ (3.8 tonf/in$^2$) and for 99.0 per cent aluminium the value is 69 N/mm$^2$ (4.5 tonf/in$^2$). The hardness of soft aluminium is 17 on the Vickers Diamond scale and rises to 40 when cold worked 80–90 per cent reduction in thickness. There is no theoretical relationship between hardness values and mechanical properties such as proof stress or elongation but, as an approximation, tensile strength values in N/mm$^2$ are about four times the hardness on the Vickers Diamond scale.

### 1.1.3.2   The effect of temperature on strength
The compressibility of aluminium increases with increase in temperature, the $dV/V_0dP$ values for 99.5 per cent aluminium being $1.45 \times 10^{-6}$ at 20 °C and $1.7 \times 10^{-6}$ at 125 °C.

The problem of brittle fracture does not arise with aluminium alloys of the f.c.c. (face-centred cube) type, and low temperatures do not lead to cleavage fractures. There is a slight increase in the notch impact strength at low temperatures [5].

When annealed for a constant time of 90 min, rapid softening of 99.99 per cent aluminium occurs in the temperature range 200–300°C but at temperatures above about 325°C, no further softening occurs (Fig. 5).

### 1.1.3.3   Creep strength
The static endurance or creep strength of material under long-time loading is defined as the stress giving rise to a steady creep rate of not more than 0.00001 per cent strain per hour. For commercially pure aluminium the creep values fall rapidly as temperatures exceed 100°C. This may be of advantage for some forming operations but renders the unalloyed metal unsuitable for applications where the metal is stressed at elevated temperatures. The drop in tensile strength is coupled with an increase in elongation value.

In considering the effect of high temperatures on mechanical properties, there has to be a clear distinction between

(a) properties determined at ambient temperature after prolonged exposure of the specimen to the elevated temperature, and
(b) the properties of the specimen when tested at the elevated temperature.

This difference is illustrated in Figs 6 and 7.

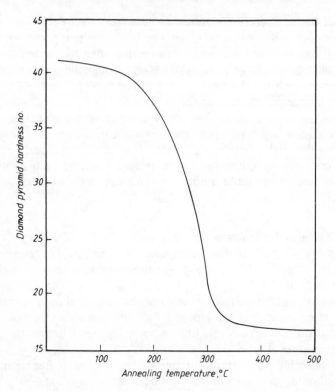

Fig. 5 — Hardness variations in cold worked (90%) 99.99% pure aluminium as a function of annealing temperature, constant annealing time 90 min (*After Van Lancker.*)

## 1.2  STANDARDS FOR ALUMINIUM AND ALUMINIUM ALLOYS

Wrought aluminium and aluminium alloy materials intended for general engineering use are specified in the British Standards 1470 to 1477 whilst those intended for more specialised use are published in a supplementary series BS 4300. Ingots and castings are covered by BS 1490. Other countries have their own national standards which usually differ from those of the British Standards Institution but work through the International Standards Organisation now ensures that the principal national standards fall within the ISO Recommendations.

These specifications, both national and international, are kept under review and, if circumstances warrant it, alterations may be made to composition limits or mechanical properties. Hence reference must be made to the latest editions of specifications to avoid the possibility of errors. In the UK, information about specifications for aluminium alloys may be obtained from

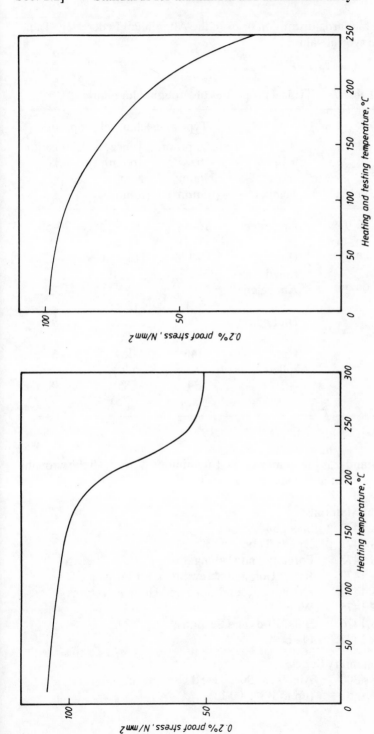

Fig. 7 — Loss of strength (0.2 proof stress) as for Fig. 6 but tested at heating temperature.

Fig. 6 — Loss of strength (0.2% proof stress) of pure aluminium after heating at 200–300°C for one year, testing temperature 20°C. (*After Von Zeerleder*).

the British Standards Institution or from the Aluminium Federation. Table 7 gives some typical values.

Table 7 — Typical properties of wrought aluminium

| Form | Purity | Temper | Typical mechanical properties | | |
|---|---|---|---|---|---|
| | | | 0.2% proof stress N/mm$^2$ (tonf/in$^2$) | Tensile strength N/mm$^2$ (tonf/in$^2$) | Elongation % |
| Sheet | 99.99% | Annealed | 16–49 (1–3) | 62–92 (4–6) | 20–50 |
| Sheet | 99.99% | Hard rolled | 77–130 (5–8.5) | 115–154 (7.5–10) | 4–15 |
| Sheet and strip | 99.2% | Annealed | 31 (2) | 85 (5.5) | 33 |
| | | Half hard | 100 (6.5) | 123 (8) | 9 |
| | | Hard rolled | 146 (9.5) | 154 (10) | 5 |
| Extrusions | | | 39 (2.5) | 85 (5.5) | 38 |

British Standards for aluminium and aluminium alloys in all the wrought forms, and as ingot and castings, are as follows:

Wrought Materials:
BS 1470    Sheet and Strip
BS 1471    Drawn Tube
BS 1472    Forgings and Forging Stock
BS 1473    Rivet, Bolt and Screw Stock for Forging
BS 1474    Extruded Round Tube and Hollow Sections
BS 1475    Wire
BS 1476    Bars, Rods and Sections
BS 1477    Plate

Supplementary Series:
BS 4300    Alloys for Specialised Use
(up to BS 4300/13)

BS 1490    Ingots and Castings

Standards for Aircraft Materials:
BSL Series as follows:

| | |
|---|---|
| 4L 16 | 99 per cent aluminium sheets (half hard) |
| 4L 17 | 99 per cent aluminium sheets (soft) |
| 6L 25 | Aluminium–copper–nickel–magnesium alloy forging stock and forgings |
| 4L 33 | Aluminium–silicon alloy ingots and castings Amendment AMD 510 1970 |

Other British Standards Relating to Aluminium:

| | |
|---|---|
| BS 3660 | Glossary of Terms used in the Wrought Aluminium Industry |
| BS 1161 | Aluminium alloy sections for general engineering. |
| BS 2613 | Aluminium alloy sections for marine purposes. |
| BS 2855 | Aluminium corrugated sheet (general). |
| BS 388 | Aluminium flake, pigments for paints. |
| BS 1683 BS 3313 | Aluminium foil for dairy product wrapping. |
| BS 2897 | Aluminium strip for electrical purposes. |
| BS 2627 BS 3988 | Aluminium wire for electrical purposes. |
| BS 6791 | Aluminoum insulated cables. |
| BS 215 BS 3242 | Aluminium overnead lines. |
| BS 1615 | Anodic oxidation coatings on aluminium. |
| BS 3987 | Anodised wrought aluminium for external architectural applications, 1974. |
| CP 143 | Sheet roof and coverings: Part 1: Aluminium corrugated and troughed (Amended PD 4346) Part 7: Aluminium |
| CP 8118 | The Structural Use of Aluminium, 1987 Code of Practice Safe Transport of Molten Aluminium by Road in Great Britain, 1974 (Association of Light Alloy Refiners) |

# REFERENCES

[1] Van Lancker, M. *Metallurgy of Aluminium Alloys*, 1967, Chapman & Hall, p. 1.
[2] Smithells, C. J., *Metals Reference Book*, 1955.

[3] Brimelow, E. I., *Aluminium in Building*, 1957, Macdonald, p. 315, Table 32.
[4] Van Lancker, M., *ibid.*, p. 41, Fig. 110, p. 191.
[5] Van Lancker, M., *ibid.*, p. 80.

**FURTHER READING**

Hydrogen in Aluminium, *Z. Metallkunde*, **52**, 1961, 682. *Aluminium (D)* June 1963, 356.
*Constitution of Binary Alloys*, M. Hansen, 1958, McGraw-Hill.
Key to Aluminium Alloys — Designations, Compositions and Trade Names of Aluminium Materials — Compiled by W. Hufnagel, Published by *Aluminium-Zentrale*, 1982, 1st edition, 232 pp.
Low-Temperature Properties of Some Aluminium Alloys — A Collection of 4 Papers from the *British Welding Journal* November 1985, pp. 499–538:
Low-Temperature Properties of Welded and Unwelded A1–5% Mg Alloy Plates, J. E. Tomlinson and D. R. Jackson.
Extruded Aluminium Alloys for Low Temperature Service — An Assessment of the Suitability of Tow Materials, R. J. Durham.
Tear Tests on Aluminium–Magnesium Alloy Plate — J. Sawkill and D. James.
Low Temperature Properties of Aluminium–Magnesium Alloys — R. E. Lismer.

For full details of British Standards and ISO Recommendations for Aluminium And Its Alloys:

British Standards Institution, 2 Park Street, London W1A 2BS

Publications Department, British Standards Institution, Linford Wood, Milton Keynes MK14 6LE.

# 2

# Occurrence and extraction of aluminium

Aluminium, unlike some of the heavy metals such as copper, gold and silver, is not found in its native state and is reducible from its chemical compounds only with difficulty. Oersted in 1825 reported having reduced aluminium chloride with potassium amalgam and later Wöhler used metallic potassium as the reducing agent. The process was greatly improved by Henri Sainte Claire Deville who used the cheaper sodium in place of potassium and the stable double salt of sodium aluminium chloride in place of the hygroscopic aluminium chloride. A modified process (German patent 26,962 of 1883) produced the double salt from a reaction between alumina, sodium chloride and tar treated with chlorine. The double salt was then reduced with magnesium. This proved too costly and in its place cryolite ($Na_3AlF_6$) was reduced by magnesium, viz.

$$2Na_3AlF_6 + 3Mg = 2Al + 3MgF_2 + 6NaF$$

The Hall and Hèroult patents both covered the electrolysis of aluminium oxide in a bath of molten halide salts and since then alumina, extracted from bauxite, and cryolite have become the most important — virtually the sole — sources of aluminium. However, impurities in these raw materials resulted in a product having an aluminium content of 90–94 per cent, the chief impurities being 5 to 8 per cent silicon, 1 to 2 per cent iron and a few tenths of a per cent of aluminium oxide.

The introduction of the Hall–Heroult process for the electrolytic production of aluminium required alumina ($Al_2O_3$) of the highest possible purity to be used, as any impurities present pass into the aluminium metal.

Fig. 8 — World map showing aluminium ore mining areas, principally for Bauxite. 1. Jamaica, 2. USA (Alabama, Arkansas, Georgia), 3. Guyana, 4. Surinam, 5. Brazil, 6. French Guinea, 7. Ghana, 8. Southern Africa, 9. Spain, 10. France, 11. N. Ireland, 12. N. Italy, 13. Austria, 14. Germany, 15. Greece, 16. Hungary, 17. Roumania, 18. Yugoslavia, 19. USSR, 20. N. W. India, 21. Malaysia, 22. Borneo, 23 and 24. Australia, 25. Greenland (for Cryolite).

## 2.1 OCCURRENCE [1]

Aluminium in combination constitutes almost 8 per cent of the earth's crust and is the most abundant of metals. It is an essential constituent of the clay minerals and of a large number of important silicates such as the felspars, micas, sillimanite, etc. The chief industrial sources of aluminium and its compounds are bauxite and to a lesser extent cryolite ($Na_3AlF_6$), alunite ($KAl_3(SO_4)_2(OH)_6$), leucite ($KAlSi_2O_6$) and alum shales. Such industrial minerals include potter's clay, china clay (kaolin), fuller's earth, garnet (the hydrated oxide), mica ore, silicates, white or near-white aluminium oxides which occur as bauxite, corundum and emery (non-hydrated).

Bauxite is essentially hydrated aluminium oxide, $Al_3O_32H_2O$, but is generally associated with impurities of iron oxide, phosphorus compounds and titania. Some material called bauxite has the composition of diaspore hydrous aluminium oxide, $Al_2O_2(OH)_2$ or $Al_2O_3H_2O$ (Table 8). The common form of bauxite is as amorphous earthy granular or pistolic masses having a colour ranging through dirty white, greyish, brown, yellow or reddish brown, averaging 60 per cent alumina against roughly 40 per cent in good kaolin or clay.

Bauxite results from the decay and weathering of aluminium-bearing rocks, often igneous but not necessarily so, under tropical conditions and it may form residual deposits replacing the original rock, or it may be transported from its place of origin and form deposits elsewhere. It occurs, for example, in France, in pockets in Cretaceous limestone and is the result of pre-Tertiary tropical weathering; other important deposits occur in N. America, particularly Alabama, Arkansas and Georgia; South America in Brazil, Guyana and Surinam; North Africa in French Guinea and Ghana; South African Republic; Austria; S.E. Australia; Borneo; Germany; N.W. India; W. Indies (in Jamaica); N. Ireland; N. Italy; Malaysia; Spain; USSR; and in Yugoslavia, as shown on the world map (Fig. 8).

Generally the bauxite is extracted by open-cast working methods. It is then washed and screened to remove extraneous dirt. The lumps of washed ore may be treated locally at the place of mining for the production of alumina, or they may be transported to more distant locations for treatment by the Bayer process. The basic production flow line is indicated in Fig. 9.

## 2.2 THE BAYER PROCESS

The washed ore is crushed to a fine powder before chemical treatment for the separation of alumina from the impurities, the chief of which are iron oxide, silica and titanium oxide. The Bayer process, which dates back to 1892, is still widely used for this process of beneficiation, as illustrated in Fig. 10.

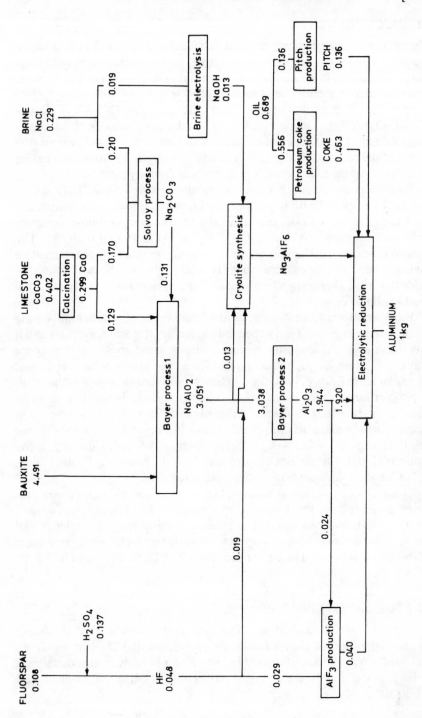

Fig. 9 — Basic principles of aluminium production. (*After Hancock.*)

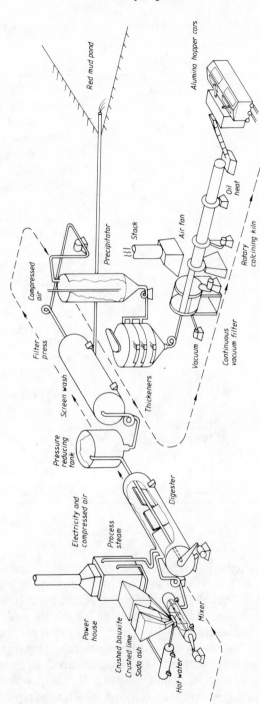

Fig. 10 — Schematic flow of materials — bauxite to alumina.

**Table 8** — Forms of alumina (Alumina exists in the $\alpha$, $\beta$, and $\gamma$ forms)

| Oxide | Properties | Structure |
|---|---|---|
| $Al_2O_3$ | Amphoteric | $\alpha$ form. Corundum inactive, high temperature form. Oxide ions c.c.p. with aluminium ions distributed regularly in octahedral sites. It is rhombohedral, density 3.96. |
| | | $\beta$ form. Hexagonal structure, density 3.31 |
| | | $\gamma$ form. Low temperature form, more reactive. Metal ions arranged randomly over octahedral and tetrahedral sites of a cubic spinel, density 3.42. |
| | | Transforms to $\alpha$ at 750°C |
| Alumina also exists in gem forms | | |
| | White sapphire | $Al_2O_3$ |
| | Ruby | $Al_2O_3$ + traces of $Cr^3$ |
| | Blue sapphire | $Al_2O_3$ + traces of $Fe^{2+}$, $Fe^{3+}$ or $Ti^{4+}$ |

The crushed bauxite is fed into large steel digester vessels containing caustic soda (NaOH) at a concentration of 45° Be which is heated to a temperature of 165°C at a pressure of about 6 atmospheres (608 kNm²) [2]. The $Al_2O_3$ is taken into solution as sodium aluminate and the silica is converted into sodium alumino-silicate, most of which is precipitated in the residual 'red mud', but traces still remain in solution. During treatment in the digester the various iron hydroxides in bauxite are dehydrated to form $\alpha$ $Fe_2O_3$ solid solution (containing up to 10 per cent $Al_2O_3$) which crystallises out to form the major constituent in 'red mud' which has a typical analysis as follows:

       56–60% $Fe_2O_3$
       16–18% $Al_2O_3$
       2–3% CaO
       0.5–1.5% MgO
       0.4–6% $Na_2O$
       6–8% $TiO_2$

The mud is diluted and sent for filter pressing. It is then pumped away as a slurry to be dumped into large man-made lagoons, and it is noteworthy that this by-product has still not found any commercial use.

After filtration the solution of sodium aluminate contains a little sodium alumino-silicate together with a colloidal suspension of iron oxide and alumina, capable of passing through the safety filters.

The filtrate is pumped into tall cylindrical steel tanks, illustrated typically in Fig. 11, and allowed to cool. The $Al(OH)_3$ is precipitated by inoculating

Fig. 11 — Two typical decomposers of which more than 150 were used in the Bayer process operated by British Aluminium Company Ltd at their Burntisland works. (*Courtesy of British Alcan.*)

the filtered solution with seed crystals of alumina and agitating. The precipitate is removed by filtration and is calcined at temperatures of 1200 to 1300 °C in long rotary kiln furnaces to drive off water and leave alumina as a white powder. Fig. 12 shows a typical calciner in which the reaction is

$$2Al(OH)_3 = 2Al_2O_3 + 3H_2O$$

The filtrate is concentrated and treated again, or may be used as one of the components for the manufacture of synthetic cryolite (see later).

Fig. 12 — Rotary calciner as used in the last operation in the Bayer process (*Courtesy of British Alcan.*)

## 2.3   THE PECHINEY SYSTEM [2]

The Pechiney system is an alternative to the Bayer process for the production of alumina. In the Pechiney process the bauxite is mixed with sodium carbonate and heated to 1300 °C to produce a mixture of $3Na_2O.Al_2O_3$, $Na_2O.Fe_2O_3$ and a small amount of sodium silicate. On treatment with water the aluminate and sodium silicate are dissolved, while the $Na_2O.Fe_2O_3$ is hydrolysed to $\gamma Fe_2O_3.H_2O + 2NaOH$; the $\gamma$-hydrate partially dehydrates and precipitates with some $Al_2O_3$ in solid solution, thus eliminating most of the iron. After the removal of the $Fe_2O_3$ precipitate, $Al_2O_3.3H_2O$ is precipitated by treatment of the filtrate with carbon dioxide. The hydrolysed alumina is then calcined.

## 2.4  STRUCTURE OF ALUMINA

There is no evidence for a free $M^{3+}$ ion, either in the solid or in the solution of compounds of aluminium or other members of the boron group of $ns^2np^1$ elements, boron, aluminium, gallium, iridium, and thallium. A number of solids, especially fluorides and oxides, are high melting and strongly bonded, but as the bonds are intermediate ionic and covalent, the stabilities of the solids are due to the formation of giant molecules with uniform bonding, as indicated in Figs 13 and 14 [3].

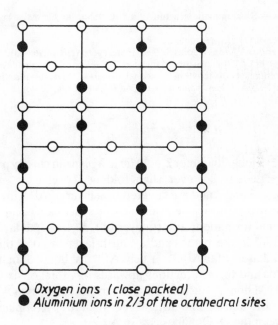

O  *Oxygen ions (close packed)*
● *Aluminium ions in 2/3 of the octahedral sites*

Fig. 13 — Corundum (elevation).

The crystal lattice of $\alpha Al_2O_3$ contains no discrete molecules of $Al_2O_3$ [4]. The aluminium and oxygen valency electrons are bonded in a complex manner by covalent homopolar bonding; there are no atoms present, only $Al^{3+}$ and $O^{2-}$ ions.

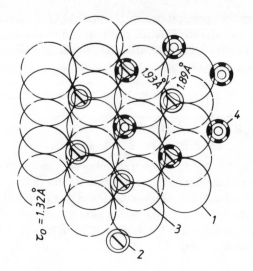

Fig. 14 — Diagrammatic structure of $\alpha Al_2O_3$, showing successive layers of $O^{2-}$ and $Al^{2+}$ ions:

1. layer of $O^{2-}$ ions (radius 1.32 Å);
2. layer of $Al^{3+}$ ions (radius 0.57 Å) in hexagons (cross-hatched);
3. layer of $O^{2-}$ (12 ions turned through 180° relative to the first layer of 12);
4. another layer of Al ions in hexagonal formation. Al–O distances in $AlO_6$ are 1.89 and 1.93 Å. (*From Van Lancker, Thesis, Paris Univ.* 1954.)

The $O^{2-}$ ions (diameter 2.6 Å) lie in a plane in closely packed hexagonal formation; there is, however, slight folding. If twelve $O^{2-}$ ions are placed in rows of 3, 4, 3 and 2 and are covered by a layer of $Al^{3+}$ ions (diameter 1.14 Å) so that two-thirds of the available free spaces are occupied by $Al^{3+}$ ions in hexagonal formation, a third layer of $O^{2-}$ ions like layer 1, but rotated through 180° in the plane, and a fourth layer of $Al^{3+}$ this completes the structural bases of $\alpha Al_2O_3$. Each $Al^{3+}$ is bonded in three directions downwards and three directions upwards to a total of six $O^{2-}$ ions. There are no Al–Al bonds. Each $O^{2-}$ ion is surrounded by two $Al^{3+}$ ions below it and two more above. The distance between the Al–O planes is thus small; that between the O–O planes is 2.16 Å.

$$\alpha Al_2O_3 = \alpha[Al^{3+} - O_{6/4}^{2-}]_\infty = \alpha[Al^{3+} - O_{(3+3)/4}^{2-}]_\infty$$

6 and 4 being the co-ordination numbers of Al and O respectively and $\infty$ indicating all directions in accordance with the pseudo-octahedral rhombohedral type.

There are no discrete $AlO_6$ octahedra in $\alpha Al_2O_3$ in contrast with $Na_3AlF_6$ which does contain discrete $AlF_6$ octahedra.

Many metals may be obtained by reaction of the element's oxide with carbon and the question may be asked as to why this cannot be achieved in the case of aluminium.

The process of extraction by carbon reduction [5] depends on the difference in the two oxidation reactions of the type

$$M + \frac{x}{2} O_2 \rightarrow MO_x \qquad (1)$$

Reactions which evolve free energy tend to occur spontaneously so that the equilibrium between two elements and their oxides or corresponding equations for different oxide stoichiometries

$$MO + M_1 \rightleftharpoons M_1O + M \qquad (2)$$

will favour that oxide whose free energy of formation (equation (1)) is most negative.

The free energy change $\Delta G$ is separable into two components, $\Delta H$ and the energy involved in the change of entropy $T\Delta S$, where $T$ is the absolute temperature.

$$\Delta G = \Delta H - T\Delta S$$

In the formation of a metal oxide according to equation (1) the heat change is usually favourable but, as the reaction uses up a gaseous component (oxygen), which has a relatively large entropy, the entropy term is unfavourable and this energy increases with increasing $T$. As a result the free energy change for metal oxide formation in equation (1) falls off with rising temperature in a broadly similar way for any metal as shown in Fig. 15.

Metal oxides can be divided into two classes: firstly, those which are intrinsically unstable at normal temperatures, such as gold oxide, or at accessible temperatures such as silver oxide or mercury oxide which may be extracted by simple heat treatment; secondly, those which contain the elements with a favourable energy of oxide formation at any economically accessible temperature. The extraction of elements of this second class involves separation by chemical reduction or by electrolytic reduction.

Any metal will reduce the oxide of any other metal which lies above it in Fig. 15, according to equation (2), as the net change in the free energy will be negative (i.e. favourable) by an amount equal to the distance between the two curves at the appropriate temperature.

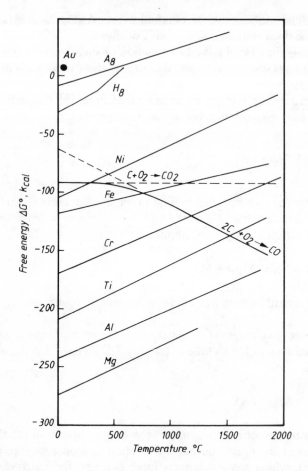

Fig. 15 — The variation with temperature of the free energy of metal oxide formation
(according to equation (1)) and for carbon with oxygen.

The formation of a metal oxide thus involves a free energy change which
becomes less negative with increasing temperature because the oxidation
proceeds with the consumption of a gaseous component with a corresponding loss of entropy. The formation of carbon oxides is quite different. The
main reactions between carbon and oxygen at higher temperatures are

$$2C + O_2 \rightarrow 2CO$$

in which an excess of one mole of gas is produced for each mole of oxygen
used. This reaction therefore involves a positive entropy change and its free

energy becomes more negative as the temperature rises. At temperatures below 500°C the main reaction is

$$C + O_2 = CO_2$$

where there is no overall change in the amount of gaseous reactant. The entropy change is thus small and the free energy of this reaction is almost independent of temperature. The total free energy curve falls as $T$ rises and will eventually cross every curve of the second class of elements mentioned earlier and shown in Fig. 15 for the free metal oxide formation. It follows that, in principle, carbon may be used to reduce any metal oxide if a high enough temperature can be attained. In practice, temperatures which would be high enough to allow carbon to reduce the more stable oxides such as $TiO_2$ or $Al_2O_3$ are not accessible economically on a large scale.

## 2.5  CRYOLITE

Cryolite (the double sodium aluminium fluoride) occurs in a pegmatite vein (coarse-grained quartz–felspar rocks formed by the consolidation of the last portion of the magma of igneous rock) in granite, at Evigtuk in West Greenland, associated with galena, blende, siderite, fluor and other minerals [6]. This deposit was the main source of cryolite for the industry during the early years of the twentieth century but increasing demand has led to the development of synthetic cryolite.

The natural rock has a cryolite ($Na_3AlF_6$) content averaging about 78 per cent, with $Fe_2O_3$ 5–12 per cent and $SiO_2$ 1.8–5 per cent.

The quarried rock is hand-sorted before being crushed to a size between 3 and 6 mm and dried in a rotary kiln. It is then fed onto an endless belt magnetic separator, the magnets on a second belt moving at right angles to the main burden carrying belt, removing the magnetic material and discharging it into hoppers alongside the belt. Removal of the iron oxide yields a product having a cryolite content of 96 per cent. This material is then passed over a Wiffler table — an inclined slatted vibrating table which makes use of differences in the density of cryolite and its impurities to separate further quantities of iron oxide and the removal of galena.

The cryolite-rich material is then mixed with soda ash and crushed before being passed through flotation cells which are used to remove more impurities from the mineral. This process depends upon differences in the wetting characteristics of the various constituents and on the ability of air agitation to produce a froth to carry specific constituents, such as iron oxide, to the surface. Small quantities of a frothing agent such as oleic acid are added to the stream of water and crushed material to lower the surface tension and produce a foam when air agitation is applied. A patent heteropolar agent applies selectively a water repellent film to the impurities, making them hydrophobic so that they are collected by the air bubbles and are floated to the surface. The material which is not wetted, during passage through a series of flotation tanks, is passed into a thickener, filtered and dried. This yields 98.5–99.8 per cent available cryolite.

Natural cryolite is increasingly being replaced by synthetic fluorides of optimum composition for the electrolysis conditions in use. Synthetic cryolite can be produced by passing fluorine gas through a solution of sodium aluminate and spent liquor from the seeding tanks in proportions to give one part aluminium to three parts sodium. The cryolite is precipitated by bubbling carbon dioxide through the treated solution. Analysis of cryolite from this process shows $Al_2O_3$ up to 10 per cent and about 3 per cent water.

Fig. 16 shows the structure of $\alpha Na_3AlF_6$ and its allotropic modification

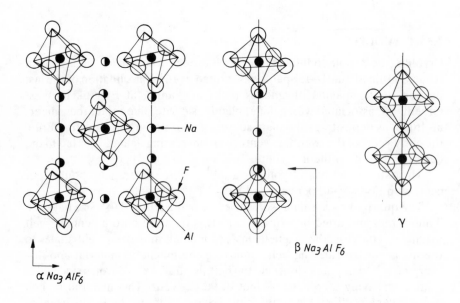

Fig. 16 — Structure of cryolite ($\alpha$, $\beta$, $\gamma$). (*After Van Lancker.*)

which is stable above 838 K (565°C) [7]. The equilibrium diagram, Fig. 17, for $NaF.AlF_3$, shows that at 885°C a eutectic is formed between NaF and $AlF_3$ at about 75 mol % NaF to give $Na_3AlF_6$.

The entropy of fusion for cryolite is 12.96 cal/°C mol. This is nearly four times the entropy value of 3.07 cal/°C ion required for the fusion of other halides. Thus, the ions present in molten cryolite $Na_3AlF_6$ should at first

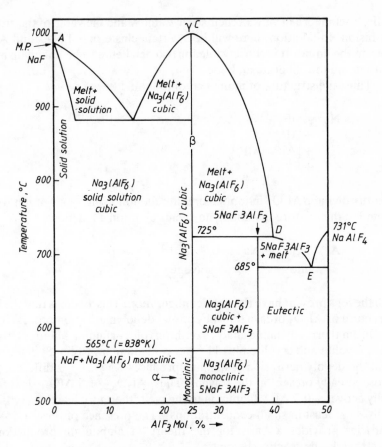

Fig. 17 — Equilibrium diagram of cryolite. (*After Van Lancker.*)

sight be three ions of $Na^+$ and one of $Al_6^{3-}$ but more information is required before the structure of molten cryolite can be fully understood.

Thermal analysis of the β modification of cryolite can be written:

$$\beta Na_3AlF_6 = \left[ Al \begin{array}{c} {}^{F_3} \\ {}_{F_3} \end{array} - Na_{8/4} \right]_\infty$$

8 and 4 being the co-ordination numbers of the $AlF_6$ and Na respectively and ∞ indicating all directions in accordance with the octahedral structure.

The F–F distances 2.53 to 2.50 Å are within the $AlF_6$ octahedron and the tetrahedron orientations are corrected above 830 K. At 1009°C, the $\beta Na_3AlF_6$ fuses; the $Na^+$ ions are retained, and so are the majority of the

$AlF_6^{3+}$ octahedra in view of their steric stability and the value of the entropy of fusion. In addition there will be a certain clustering of $Na^+$ and $AlF_6^{3-}$ ions in the liquid. It is conceivable for two octahedra to be linked up at one corner and thus liberate an F ion.

Thus, the structure of molten cryolite should be

$$Na^+, AlF_6^{3-}$$
$$+ Al^{3+}$$
$$+ [AlF_5–AlF_6]$$
$$+ F^-$$

On dissolving $\alpha Al_2O_3$ in molten cryolite the following types of ion will be found in the mixture, assuming no chemical or ionic reactions.

$$Al^{3-}, O^{2-}; \quad Na^+, Al^{3+}, AlF_6^{3-}, F^-$$
$$\text{solute} \qquad\qquad \text{solvent}$$

All the ions present have the neon configuration. This assumes that the layer structure of $Al_2O_3$ breaks down by a total depolymerisation process.

From thermodynamic analysis of the cryolite–alumina liquidus between pure cryolite and cryolite plus 11.5 per cent alumina, Rolin [8] determined that the depolymerisation of alumina produces 1.5 particles different from those already present in the solvent. Thus, $Al_2O_3 \rightarrow 1.5\ AlO_2^-$ (cryoscopically active) $+ 0.5\ Al^{3+}$ (common to the melt, therefore inactive cryoscopically — at least in small concentrations). The presence of these 1.5 $AlO_2^-$ particles provides a basis for the theory of aluminium production by electrolysis, the reaction being

$$2Al_2O_3 \rightarrow 3AlO_2^- + Al^{3+}$$

Electrolysis of the production solutions obeys the simple formula

$$Al^{3+} + 3e^- \rightarrow Al$$

$Al^{3+}$ being one of the ions produced by the depolymerisation of $2Al_2O_3$. The discharge of $O^{2-}$ or $AlO^{2-}$ from the same depolymerisation reaction accounts for the formation of $CO_2.AlO_2^+$; $2AlO_2 \rightarrow Al_2O_3 + \frac{1}{2}O_2$; $O_2 + C \rightarrow xCO$ and $yCO_2$, the ratio $x/y$ depending on the shape of the anode, for example.

The gradual accumulation of $AlF_3$ round the anode in electrolytic cells is a result of the mobility of $AlF_6^{3-}$ which migrates towards the anode under the influence of the current and is not discharged.

## 2.6 PRODUCTION BATHS

Electrolytic baths for the reduction of alumina to aluminium are often neutral (2 to 8 per cent $Al_2O_3$) and eutectic temperature is 935°C. Others are acid (up to 10 per cent $AlF_3$ added) basic (NaF added) or complex (e.g. $CaF_2$ added) [9].

The effects of these additions on the electrical conductivity are summarised in Table 9.

**Table 9** — The effect on electrical conductivity of additions to cryolite production baths

| Composition of bath | Conductivity mho/cm |
| --- | --- |
| 100% cryolite | 2.36 |
| 90% cryolite 10% $CaF_2$ | 2.53 |
| 90% cryolite 10% $AlF_3$ | 2.27 at 960°C |
| 93% cryolite 7% $Al_2O_3$ | 2.13–2.17 at 960°C |

### 2.6.1 Power requirements [10]

The theoretical quantity of electricity required for the production of 1 g of aluminium is 2.975 Ah (2.975 kAh/kgAl), but the Faraday yield is reduced by the burning of Al in $CO_2$ at the anode, viz.

$$2Al + 3CO_2 = Al_2O_3 + 3CO$$

Thus, for each mole of CO gas, two Faradays have been lost. The normal yield is $n = 85$ per cent for baths in good condition, with a maximum of about 90 per cent of the theoretical quantity; the yield is increased if the concentration of $Al_2O_3$ is kept constant.

Rolin [11] and Mergault [12] have established that the minimum voltage for the decomposition of the electrolyte, above which there is detectable electrolysis, is $e = 1.07$ V at 1070°C; 1.01 V at 980°C with current densities of 0.08 $A/cm^2$ at the anode and 0.25 $A/cm^3$ at the cathode, both electrodes being graphite. The value 1.07 V is almost independent of $Al_2O_3$ content.

If there are insufficient $O^{2-}$ or $AlO_2^-$ ions in solution, $F^-$ ions are discharged at the anode and $F_2$, $CF_4$ or CO are formed, this being known as the anode effect.

The effective electrolysis voltage is around 1.6 V, the exact value depending upon the current density, size and type of electrodes, temperature, composition of the bath (electrolyte) and other factors.

Considering the overall reaction leading to the deposition of aluminium together with the formation of CO and $CO_2$ and adding to this sensible heat of the products Ferand, L. [13] calculated that the overall power consumption is between 5.4 and 6.4 kWh per kg of aluminium. This figure allows for anode depolarisation reactions, as against the theoretical figure of 2.975 × 1.07 kWh discussed above.

Production cells operate at about 5 V. The difference between this value and the 1.6 V quoted for effective electrolysis is accounted for by cathodic and anodic loss, 0.3, and 0.25 V; inter-polar voltage drop, 1.6 V (resistivity drop); effective voltage, 1.7 V; loss between cells, 0.1 V; loss due to $Al_2O_3$ deconcentration, 0.1 V: Total 4.05 V. The power consumption per kg of aluminium extracted is between 15 and 18 kWh. Thermal losses through the anodes, the electolyte and the cathode, as well as that resulting from the gas reactions, all contribute to the low yield.

### 2.6.2   Bath construction and operation
Electrolysis cells may be divided into two main types:

(i)    those using prefired electrodes, and
(ii)   those having electrodes fired in the bath, i.e. Soderberg electrodes. (See Figs 18 and 19.)

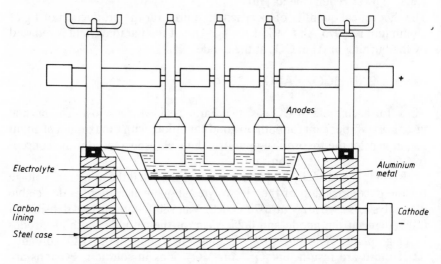

Fig. 18 — Diagrammatic section through aluminium electrolytic cell using Soderberg electrodes (Noguères plant).

The sizes of reduction cells are usually expressed in terms of the electrical capacity of the individual cell in kA. Cells with prebaked anodes may be

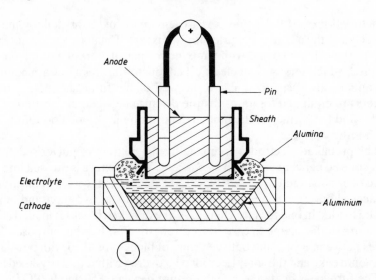

Fig. 19 — Diagram to show the layout of the electrolytic cell using the Soderberg electrodes at Noguères plant. (*After Van Lancker*.)

rated as having a capacity of 20 to 105 kA whereas a Soderberg cell may have a capacity of up to 150 kA. (See Table 10.)

**Table 10** — Operating data for electrolytic cells of different types and capacities

| Type of cell | Prebaked anodes | | Soderberg anodes |
|---|---|---|---|
| Capacity of unit (kA) | 85–100 | | Up to 150 |
| Number of anodes | 5 | 5 | |
| Width of anodes (cm) | 71 | 71 | |
| Length of anodes (cm) | 320 | 320 | |
| Total anode area (cm²) | 111,000 | 111,000 | |
| Total cathode area (cm²) | 176,000 | 176,000 | |
| Current (kA) | 82 | 105 | 92 |
| Anode density (A/cm²) | 0.72 | 0.93 | |
| Thermal voltage (V) | 4.35 | 4.95 | 5 |
| Output per unit per day (kg) | 594 | 760 | |
| Specific power consumption (kWh/kgAl) | 14.4 | 16.4 | 14.5 |

The construction of a reduction cell consists of an outer shell of steel plate suitably reinforced. The rectangular shell is lined with insulating and refractory bricks. Metal conductors are laid on top of the brick floor and are connected to the negative electric supply busbars. The floor, conductors and walls of the cell are coated with a mixture of petroleum coke or pitch coke and pitch which serves as an electrical conductor having a low ash content. The anodes are suspended above the well of the furnace. The electrode carriers are equipped for adjusting the distance between the bottom of the anodes and the surface of the molten deposit on the carbon floor of the cell which is the cathode.

The prebaked electrodes are made from a mixture of pulverised petroleum coke 80 per cent and pitch 20 per cent. This mixture is pressed into the shape of anodes and the 'green' blocks are stacked in layers in open-topped furnaces about 27 m long and 3 m wide. As many as two thousand blocks may be loaded in one furnace charge. A line of baked blocks is loaded along the centre of the charge to carry the current used for heating purposes. The spaces between layers and between rows of blocks are filled with powdered petroleum coke and the completed load is covered with refractory-lined lids.

The furnace is heated to a temperature between 850 and 900°C for 7–8 days and this is followed by cooling for a further 20 days.

The rods used for connecting the anodes to the busbars are placed in holes made in the tops of the blocks during the pressing operation and are secured in position by pouring cast iron into the annular space thus formed.

Soderberg electrodes can be made in any desired size by filling a mixture of pulverised petroleum coke 70–75 per cent and pitch 30–35 per cent in the form of pellets or a plastic mass into a sheet aluminium mantle which can be fed down into the electrolyte as the lower portion of the carbon mixture is first of all baked and then burnt during the electrolytic process.

Both methods of electrode production yield copious volumes of fumes which are exhausted and treated for the recovery of the tar content. The fumes and dust from the reduction pots also contain fluorine and fluorides which are also recovered for reprocessing [14].

Reduction units are also equipped for the automatic control of electrolysis conditions, voltage (or current), equal voltage distribution between the anodes, temperature (a function of voltage across the terminals, total current and yield) and alumina feed.

The d.c. power supply is taken from rectifiers (mercury vapour, germanium or silicon-type rectifiers). Reduction cells are normally installed in long lines along each side of single-span buildings equipped with overhead cranes for servicing the units. Each bay will house between 100 and 120 cells connected in series.

As the starting up of an electrolytic cell, from cold, is a time consuming and costly operation, it is important to avoid any prolonged interruption in the power supply. Normally, once in operation, cells are run continuously for years until taken out for rebuilding or some other major maintenance work. At any one time a small number of cells may be taken out for this purpose.

A most important tool in the operation of electrolytic cells is the pot crust breaker — a pneumatic chisel actuated at the end of a tilted rocking beam. This is used to break the hard solidified crust which forms over the electrolyte. This crust has to be broken to allow alumina and other additions to enter the molten electrolyte.

At intervals dependent upon the size of the units, the aluminium is siphoned off into refractory-lined ladles (Fig. 20) and transferred into large holding furnaces of several tens of thousand kilograms capacity where the yields of several reduction units are mixed. The weight of metal from each

Fig. 20 — Molten aluminium being syphoned from electrolytic reduction cells. The extent of this 'pot room' is clear. (*Courtesy British Alcan.*)

unit is recorded and samples are taken for spectrographic analysis. This monitoring permits the early detection of any abnormality in cell performance and also allows close control of the composition of the blended metal. Frequent checks at regular intervals are also carried out on samples of electrolyte to ensure that the correct balance in the amounts of the various components is maintained.

The metal from the reduction pots is at a temperature in excess of 900°C.

It is allowed to stand in the holding furnace until it has cooled to a temperature between 700 and 730°C suitable for casting into ingot forms for remelting or forms suitable for processing into sheet, plate, extrusions, rolled rod or other wrought products.

During the holding period, inclusions and heavy particles sink to the bottom of the furnace, and entrapped electrolyte rises to the surface to be skimmed off.

Both alumina and cryolite contain $\alpha Fe_2O_3$ and $SiO_2$ as impurities. The measured decomposition voltage for $Fe_2O_3$ is 0.23 V against 1.07 V for $Al_2O_3$. The cathode deposit using graphite electrodes is iron but aluminium deposited during electrolysis of $Al_2O_3$ will dissolve any iron which may be formed. The other main source of iron in aluminium is the presence of $Fe_2O_3$ in the electrodes.

The decomposition voltage for $SiO_2$ is 1.07 V, as for $Al_2O_3$. Mergault reported that on electrolysis of $SiO_2$ in $NaAlF_6$ the only product is Al and no silicon. However, when aluminium is present on the cathode there is a further reaction:

$$3SiO_2 + 4Al = 2Al_2O_3 + 3Si$$

The silicon produced in this way will dissolve in the aluminium.

Titanium dioxide, another impurity in alumina, is not reduced electrolytically because of the complex nature of its oxide forms but it reacts with molten aluminium to form Ti (or $TiO_x$). Titanium can form the carbide, TiC, by reaction with colloidal carbon held in suspension in molten aluminium. This TiC is very stable and acts as a grain-refining agent in aluminium and its alloys. Any titanium which does not react to form the carbide goes into solution in aluminium and has an adverse effect on the electrical conductivity of the metal. Other elements present in electrolytically reduced aluminium may include manganese, vanadium, gallium, zinc, copper, magnesium, sodium, calcium, lithium and phosphorus in amounts ranging from a few tenths of one per cent for elements such as zinc, copper and magnesium to a few hundreths of one per cent or less for the others. Hydrogen will also be present in the molten metal, and although the amount will decrease with a fall in the temperature, a degassing treatment is required to reduce the gas content to an acceptable level. This is dealt with in Section 6.5 on melting and casting.

The purity of the metal from the pot lines will generally be in the range 99.5–99.8 per cent aluminium. When starting up new reduction units, the iron content may be high, yielding metal of only 98–99 per cent purity.

Mention has already been made of the fact that one of the earliest uses of reduction cells was to produce aluminium alloys. This practice may still be used for the production of high melting point master alloys such as Al–12 per cent Mn, advantage being taken of the high temperature at which reduction cells operate. For Al–Mn master alloy, $MnO_2$ additions are made to the reduction cell. For Al–2 per cent Cr, additions of $Cr_2O_3$ are made to the cell and for Al–0.7 per cent Ti, additions of $TiO_2$. For other alloys such as Al–25 per cent Cu, Al–5 per cent Fe or Al–12 per cent Si, additions of 100 per cent alloying element are made to the molten metal in the holding furnace.

## 2.7 POSSIBLE ALTERNATIVE REDUCTION PROCESSES

Very high electrical power required by the Hall–Hèroult process led to continual efforts to reduce the necessary energy both by improving the efficiency of pot room and by seeking alternative means of extraction which, of course, had to take account of the fundamental factors already discussed. Gradual improvements in the efficiency of the cells have brought about reductions in power of 20–25 per cent and many alternative fused salt mixtures have been tried, such as sodium aluminate. Another approach has been to electrolyse mixtures of salts of the alkaline earths resulting in alloys from which aluminium could be subsequently obtained.

One of the most interesting alternatives was the extraction process based on the formation of aluminium subhalides, particularly the chloride, followed by its subsequent reduction to yield the metal. Despite much research and development expenditure in this area, the Hall–Hèroult process remains after more than a hundred years the predominant — and almost the only — method used commercially.

*The Gross process* [15]
In 1939, Willmore based a patent on his observation that in the presence of halides, aluminium vaporises at considerably lower temperatures than that at which it normally evaporates.

As a result of work by Gross, at the Fulmer Research Institute, it was established that the Willmore process involved the subhalide, aluminium mono chloride (AlCl). By increasing the reaction temperature to more than 900°C and decreasing the $AlCl_3$ pressure to less than 10 mm, the isolation of the metal was possible, and theoretical considerations of the energy levels of the three-valency electron of the aluminium atom formed the basis of a series of patents.†

*The Toth process* [15]
This process did not necessarily use bauxite, but ores including clay were first calcined and then subjected to chlorination. The resulting aluminium

---

† British Patent 582579 (1946), British Patent 66869, etc., US Patent 2470306 (1949) were among the patents taken out.

chloride, after purification, was reduced by manganese, the aluminium appearing as powder. The chlorine and manganese were recovered for re-use and the main by-product was fine silica sand and the chlorides of other metals.

The development of the process on a commercial basis did not materialise.

*Aluminate reduction* [15]
Bonnier has described the electrolysing of sodium aluminate and, more particularly, aluminate baths containing calcium, barium or magnesium to give aluminium–calcium, aluminium barium and aluminium–magnesium alloys of various compositions. The baths could be operated at significantly lower temperatures, some of them down to 700–800°C, but the process has not proved practicable.

**REFERENCES**
[1] Read, H. H., *Rutley's Elements of Mineralogy*, 1967, Thomas Murby & Co., George Allen & Unwin Ltd. pp. 301–310.
[2] Van Lancker, M., *Metallurgy of Aluminium Alloys*, 1967, Chapman & Hall, p. 35.
[3] Mackay, K. M. and Mackay, R. A., *Introduction to Modern Inorganic Chemistry*, Interdent Books, London, p. 213.
[4] Van Lancker, M., *Metallurgy of Aluminium Alloys*, 1967, Chapman & Hall, p. 18.
[5] Mackay, K. M. and Mackay, R. A., *ibid.*, pp. 105–106.
[6] Read, H. H., *ibid.*, p. 308.
[7] Van Lancker, M., *ibid.*, p. 20.
[8] Rolin, M., *Bull. Soc. Chim. Fr.*, Feb. 1960, p. 1201.
[9] Van Lancker, M., *ibid.*, p. 27.
[10] Van Lancker, M., *ibid.*, p. 29.
[11] Rolin, M., *Bull. Soc. Fr. Electr.*, Oct. 1956.
[12] Mergault, P., *Bull. Soc. Fr. Electr.*, Oct. 1956, pp. 670–699, July 1958.
[13] Ferand, L. 'Histoire de la Science et des Techniques de l'Aluminium', Vol. 1, Le Passe; Vol. 11, Le Present. *J. Fr. Electrique*, Paris, 1961.
[14] *Engineering Aspects of Pollution Control in the Metals Industries*, Proceedings of Conference, Nov. 1974, The Metals Society:
Budd, M. K., p. 144 *ibid*.
Wood, J., p. 152 *ibid*.
King, F., p. 159 *ibid*.
Miller, J., p. 165 *ibid*.
[15] West, E. G., *Aluminium Industry*, Jan. 1986, **5**, No. 1, page 8.

# 3

# Refining of aluminium

Aluminium produced by the electrolysis of alumina in a molten bath of cryolite normally has a purity between 99.5 and 99.7 per cent aluminium, the chief impurities being iron and silicon. By attention to the selection of the raw materials and to the method of operation of the reduction plant, aluminium of 99.8–99.9 per cent purity is produced on a commercial scale. For many applications where metal with a low impurity content is required, this higher level of purity is adequate. The iron and silicon contents of this metal vary according to the source of the raw materials and the production process employed.

To obtain metal of purity higher than 99.9 per cent aluminium, the metal from the alumina–cryolite reduction cells has to be subjected to further processing. Electrolytic refining processes have been developed which are a type of electrolytic filtration by means of which metal having a purity of more than 99.99 per cent aluminium can be produced [1], the chief impurities being iron, silicon and copper in quantities totalling a few parts per hundred thousand, the content of individual elements normally being quoted in terms of parts per million (see Table 11).

The refined metal has high ductility, high electrical conductivity, excellent corrosion resistance and high specular reflectivity — all properties extremely useful for certain applications for which the cost of the extra refining operation can be justified.

Chemical methods cannot be used for the refining of aluminium because the total energies for the formation of compounds of the impurity elements such as iron, silicon and copper, with oxygen, fluorine or chlorine, are unfavourable compared with those for aluminium. For this reason, physical methods have to be employed for the removal of such elements.

**Table 11** — Analysis of refined aluminium obtained from (a) copper-rich
and (b) zinc-rich anode metal [10]

|  | (a) | (b) |
|---|---|---|
| Al | By difference 99.997% | By difference 99.996% |
| Fe | 5 ppm | 3 ppm |
| Si | 3 ppm | Not quoted |
| Cu | 20 ppm | 1.9 ppm |
| Mg | Mg+Zn 5 ppm | Not quoted |
| Zn |  | 20 ppm |
| Na |  | 1–2 ppm |
| Mn |  | 0.3 ppm |
| Cd |  | 3.5 ppm |
| Sb |  | 1.2 ppm |
| Ni |  | 2.3 ppm |
| Ce |  | 0.3 ppm |
| Sc |  | 0.4 ppm |
| Total | 33 ppm | 32.6 ppm |

One of the methods used for refining is to effect the rejection of
impurities under the influence of differences in solubility in solid and molten
aluminium respectively (the principle of zone refining). Such a process can
be operated to yield exceptionally high purity aluminium (total impurity
content 10 ppm = 0.001 per cent) from material of purity between 99.990
and 99.998 per cent aluminium. Material of this purity is required for
scientific research into the properties of extremely pure aluminium. However, on account of low diffusion rates this method of refining has been
restricted to small-scale laboratory quantities of material and has not yet
been exploited commercially.

## 3.1 ELECTROLYTIC REFINING

A method to further purify aluminium was first proposed by Hoopes in 1901
and this was improved by Betts in 1905, using a three-layer electrolytic bath.
The process was further developed by Frary and Edwards, and later by
Gadeau with others, making it possible to produce superpurity aluminium
(99.99+ per cent).

The electrolytic refining cell comprises three molten layers, as follows.

1. The lower layer is of commercially pure metal as produced by the

electrolytic reduction of alumina in a cryolite bath under carefully controlled conditions, or, more usually, an alloy of such metal, e.g. 30 per cent copper–aluminium.
2. The electrolyte, such as that developed by AIAG Neuhausen:

$$48 \, wt\% \, AlF_3, \quad 18 \, wt\% \, NaF, \quad 18 \, wt\% \, BaF_2, \quad 16 \, wt\% \, CaF_2 \ .$$

3. The purified metal.

The AIAG cell [2] operates at a temperature of 740–760°C. Aluminium from the unrefined molten alloy or metal which forms the anode is taken into solution in the electrolyte and is redeposited as pure aluminium at the cathode. The stages are

Impure metal → ionic conductor with Al ions (valency 3)
→ pure metal

The schematic arrangement in Fig. 21 shows a section through a cell

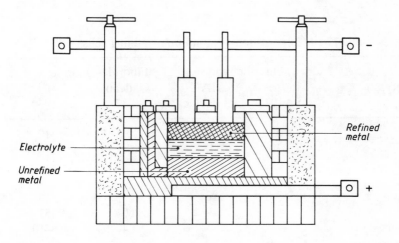

Fig. 21 — Section through a refining cell (Gadeau process).

consisting of a casing of steel plate. The walls are lined internally with 18 cm of alumina–silica insulating brick and magnesia bricks which do not conduct electricity.

The hearth blocks are of carbon and are connected to steel busbars which are carried through the casing and are rammed with a mixture of pitch and carbon powder. The prebaked graphite electrodes are attached to aluminium cross-beams by a system of clamps. These cross-beams which are connected to the cathode busbars can be moved up and down by mechanical jacks.

The metal to be purified is introduced into the bottom of the cell in either the solid or the molten state through a side channel. After the initial charge, further additions are made in the form of unalloyed aluminium to replace the metal which passes into the electrolyte and is redeposited as purified aluminium. Small additions of aluminium–copper alloy may be made from time to time to maintain a constant copper in the anode metal.

The ionic conductor is a flux in which aluminium will dissolve (forming $Al^{3+}$ or more complex ions). This flux is required to have a density which is less than that of the impure anode metal (typically about 3) and greater than that of the refined aluminium (2.3) at the operating temperature of the cell. For a fluoride–chloride bath operating at 1000 K (727°C) the density of the flux is 2.7. For other fluxes the operating temperatures and the densities of the various layers of the refining cell will depend upon the composition of the electrolyte, details of some representative compositions are contained in Table 12.

**Table 12** — Composition of aluminium-refining cells (wt%)

| Process salt | Fluoride type | | Fluoride–chloride type | |
|---|---|---|---|---|
| | Hoopes USA1922 | AIAG 1937 | Gadeau 1937 | 1960 |
| $AlF_3$ | 25–30 | 48 | 20.5 | 19 |
| NaF | 25–30 | 18 | 15.5 | 16 |
| $BaF_2$ | 30–38 | 18 | | |
| $CaF_2$ | — | 16 | | |
| $Al_2O_3$ | 0.5–3.0 | | | |
| $BaCl_2$ | — | — | 60 | 57 |
| NaCl | — | — | 4 | 6–10 |
| Impurities $CaF_2$, $Al_2O_3$ | — | — | — | 2 max |
| $MgF_2$ | | | | |
| Operating temperature | 950–1000°C | 740–760°C | | 740°C |

The depth of the electrolyte must be kept at a minimum consistent with avoiding contamination of the refined metal through contact with the unrefined metal layer. Addition of electrolyte materials is made to the central layer of the cell, through the layer of refined metal, by feeding them through a graphite ring which serves as a tundish.

The ends of the graphite rods which form the cathodes are immersed in the layer of refined aluminium which covers the electrolyte. In some designs of refining cells the graphite rods may be dispensed with and in their place molten refined aluminium in graphite-lined channels built into the top of the cell structure which may be used as the cathode. This type of arrangement facilitates insulation of the top of the cell and reduces both graphite and power consumption; it is known as the liquid cathode system.

At intervals during electrolysis the refined aluminium is syphoned off but care has to be taken to leave a layer of metal about 8 cm thick covering the flux layer to ensure good current distribution after siphoning.

A 20 kA refining cell will produce about 160 kg of refined metal per day, and since the weight of impurities removed amounts to between 3 and 5 g per kg or 0.48 and 0.80 kg per day it will be appreciated that it is a costly process.

It will be seen from Table 12 that fluoride baths may be used as alternatives to fluoride–chloride baths. The power and graphite consumptions for both types of bath are similar and the grades of metal produced are comparable. Typical consumption figures are given in Table 13.

**Table 13** — Consumption of materials per metric tonne of refined aluminium produced [16]

| | |
|---|---|
| 99.5% Al | 1025 kg |
| Copper | 5 kg (recovered from segregates) |
| $BaCl_2$ | 40 kg |
| $AlF_3$ | 30 kg |
| NaCl | 20 kg |
| Graphite cathodes and tools | 12 kg |
| Electrical power at 20 kA and 5.9 V | About 18,000 kWh |

As the concentration of impurities increases in the anode layer at the bottom of the refining cell they form segregates which crystallise in the colder areas of the side channel from where they can be removed. This metal is sorted by a liquidation method to recover the aluminium and part of the copper, and eliminates most of the iron to give metal which can be returned to the cells and to leave a residue of about 50 per cent aluminium with 8.15 per cent iron, 7.12 per cent silicon, 12–22 per cent copper, zinz, etc., that can be used as a constituent in a thermit mixture.

If it is required to produce refined metal with low copper contamination, an aluminium–40 per cent zinc alloy may be substituted for the aluminium–30% copper alloy.

Success in refining depends upon the care taken in maintaining the levels of the various layers and stabilising the operating temperature.

## 3.2  CHOICE OF ELECTROLYTE

*(a)  Fluoride baths* [3]

Mention has already been made of the need to consider density when formulating the composition of the flux to be used for the refining cell. Other properties which must be taken into acccount include stability, viscosity, volatility and electrical conductivity — see Fig. 22.

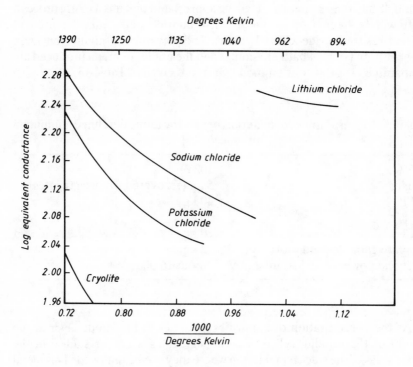

Fig. 22 — Effect of temperature on conductivity of molten salts.

Since fluorides are used in the primary reduction cells they appear to offer the first choice for the refining cells, with the proviso that the cations must not be discharged preferentially to $Al^{3+}$ on the upper cathode. Thus, the choice is narrowed to those elements with a standard free energy of formation at 1273 K (1000°C) per mole of $F_2$ (i.e. at identical discharge valencies) higher in absolute value than that of the imaginary compound $AlF_3$.

The standard free energy $\Delta G$ values at 1273 K per mole $F_2$ are

| | | | |
|---|---|---|---|
| $\frac{2}{3}AlF_3$ | — 160 kcal | $BaF_2$ | — 238 kcal |
| $2NaF$ | — 208 kcal | $CaF_2$ | — 240 kcal |
| $\frac{1}{2}MgF_2$ | — 211 kcal | $Na_3AlF_6$ | — value not known |

There is a choice of sodium, magnesium, barium and calcium, since these elements will not be discharged preferentially to aluminium provided there is sufficient $AlF_3$ present in the electrolyte. If several of these fluorides are used together, interactions can occur such as those producing $Na_3AlF_6$ and $NaMgF_3$, and the ions of the original constituents may no longer be present in the bath.

The free energies of formation of the fluorides of impurities such as iron, calcium, manganese or zinc present in the anode alloy are lower in absolute value than that for $\frac{2}{3}AlF_3$ and will not respond to the discharge of $F^-$ ions on the anode surface in preference to aluminium and therefore will not be taken into solution in the flux. On the other hand, any magnesium metal present in the anode metal will react to form $MgF_2$ and will remain in the flux layer bonded to NaF. (This assumes that Al goes into solution in the trivalent, not monovalent, form.)

## 3.3   CHLORIDE–FLUORIDE BATHS [4]

If barium chloride is added to pure molten cryolite it has been deduced that the following ions will be present:

$Al_6^3$
$Al_2F_{11}$    (two $AlF_6$ groups joined at a corner: see Fig. 15)
$Na^+$       (attracted to $F^-$)
$Al^{3+}$      (a few)
$Ba^{2+}$
$Cl^-$
$F^-$        (a few)

Mergault [5,6] showed that the decomposition voltage of cryolite between graphite electrodes is very high; the electrolysis plus anode effect leads to little aluminium and fluorine but sparking may occur. The $Ba^{2+}$ and $Na^+$ ions will be surrounded by other ions and the situation can be represented by considering the standard free energies of formation $\Delta G$ at 1000 K.

| | | | |
|---|---|---|---|
| $\frac{2}{3}AlCl_3$: | $\Delta G = -92$ kcal | $\frac{2}{3}AlF_3$: | $\Delta G = -194$ kcal |
| $2NaCl$: | $\Delta G = -153$ kcal | $2NaF$: | $\Delta G = -224$ kcal |
| $BaCl_2$: | $\Delta G = -166$ kcal | $BaF_2$: | $\Delta G = -247$ kcal |

The Ba ions will be surrounded preferentially by F ions as will Al ions. There are, however, not enough F ions available, since they are bound up in the $AlF_6$ complexes.

Thus, the electrolysis of pure molten cryolite containing barium chloride will result in the formation of aluminium in small amounts since it must come from the breakdown of stable $AlF_6$ complexes and barium at the cathode; sodium will not be formed since the Na ions are bonded, although weakly, to the fluorine in $AlF_6$ groups ($\Delta G$ for NaF is higher than for NaCl) the chlorine evolved at the anode thus derives mainly from the $BaCl_2$ and partly from the $AlCl_3$. The fact that $F_2$ is not formed at the anode can be explained by reference to the ionic composition and co-ordinations resulting from $\Delta G$ relationships. Thus, it is essential to distinguish between $F^-$ ions and the fluorine in $AlF_6^{3}$.

If free $F^-$ ions are added to the electrolyte in the form of NaF or $AlF_3$, the $Ba^{2+}$ will be surrounded by them (cf. $\Delta G$ $BaF_2$ and $BaCl_2$). Under these conditions the electrolysis will produce Al only at the cathode with no Ba, and $Cl_2$ at the anode.

If the electrolysis is stopped, chemical reaction between the chlorine and the Al–30 per cent Cu in the anode layer will produce $CuCl_2$ which may go into solution in the bath and subsequently discharge as copper at the cathode, thus ruining the refined metal.

From the point of discharge during electrolysis the $Na^{2+}$ ions are largely attracted by the $AlF_6$ complexes; the $Cl_2$ ions discharge by taking electrons from the Al atoms which go into solution. This can be represented by imaginary sets of $Al^{3+}$ ions diffusing from bottom to top of the bath.

The original Hoopes bath for the production of pure aluminium (see Table 12) had to be operated at temperatures between 950 and 1000°C (1225–1275 K), causing greater erosion of the magnesia blocks lining the cell than does the modern chloride–fluoride bath operating at 740–760°C (1013–1033 K). The magnesia blocks are also impregnated to a lesser extent and thermal insulation is improved.

The fluoride bath is more stable but has a higher electrical resistivity than the chloride–fluoride bath, with the result that lower current densities of 0.4 Acm$^2$ (fluoride) against 0.6 Acm$^2$ (chloride–fluoride) must be used to maintain the heat balance of the unit which can be regarded as a type of resistance furnace.

**3.4   ENERGY REQUIREMENTS [7]**

The theoretical amount of energy required for refining can be arrived at as follows.

(a)  The extraction of 1 g-atom of pure aluminium from the anode layer (1) of Al–30 per cent Cu requires $-\overline{\Delta G}_{Al_{(1)}}$, free molar partial energy = $+ RT\ln\alpha_{Al(1)_4}$
where $R = 1.98647$ cal/degree K g-atom, $T = 1000$ K and $\alpha_{Al(1)} = N_{Al\gamma Al}$ where $N$ is the atomic fraction of $\alpha$Al in layer (1), corresponding at 30 per cent Cu to about 86 at % Al or 0.86; $\gamma$ is less than unity and $\alpha_{Al(1)}$ is thus about 0.83.

(b)  The conversion of this g-atom of aluminium to the ionic state required $\overline{\Delta G}_2$.

(c)  The solution of this g-atom in layer (2) requires $\overline{\Delta G}_3$.

(d)  The reversal of processes (c) and (b) requires $-\overline{\Delta G}_3$ and $-\overline{\Delta G}_2$. The absolute values are the same with the signs reversed.

(e)  The introduction of 1 g-atom of aluminium into layer (3); $\Delta G_{Al(3)} = -RT\ln\alpha_{Al(1)}$.

Items (b), (c) and (d) cancel one another. Furthermore, since the refined metal is virtually at a molar fraction of unity, $\ln\alpha_{Al(1)} =$ zero, thus leaving the item $\Delta G$ for more detailed consideration.

The energy of transfer is $\overline{\Delta G}_{Al} = 23,006\, V_{Al}e$, since each g-atom of metal requires a transfer current of $V_{Al}$ Faradays and the energy required is the product of coulombs and emf, $e$ the voltage across the cell. With operating conditions as (a)

$$e = \frac{-RT\ln_{Al(1)}}{3 \times 23,066} = \frac{1.98647 \times 1000 \ln (0.83)}{3 \times 23,006}$$

$$e = \sim0.005 \text{ V}$$

In production, the operating voltage is about 5.9 V and the energy consumption is 18 kWh per kg which is equivalent to 3.06 Ah per kg. The consumption of materials is given in Table 13.

The voltage drop of 5.9 V across the cell terminals [8] includes the driving voltage required to transfer the Al = 0.005 V, about 0.35 V occurring between the busbars and the refined aluminium and 0.35 V between the anode alloy and busbars; the difference of 5.2 V is the ohmic drop across the electrolyte and interfacial drops between the three layers of the cell. Thus, it will be appreciated that whilst minor savings in energy consumption may be effected by reducing heat losses by improved insulation, very little can be done to reduce the other sources of energy loss.

Measurements by Edwards *et al.* [9] indicate that the best values for conductivities (ohm$^{-1}$-cm$^{-1}$) at 1000 °C were:

Cryolite 2.80                          Potassium chloride $2.65_2$
Sodium chloride $4.17_4$               Lithium chloride (at 700°C) 6.14

In production, the current consumption is 3000 kAh/tonne [10]

Theoretically, one Faraday (26.8 Ah) will produce

$$\frac{26.98}{3} = 8.99 \text{ g Al}$$

so that to produce 1 tonne

$$\frac{1000 \times 1000 \times 26.8}{8.99} \text{ Ah} = 2960 \text{ kAh}$$

would be required. In other words the current yield is

$$\frac{2960}{3000} = 100 = 99\% \text{ efficient.}$$

However, the energy consumption is the product of the current and the emf, which is $2960 \times 0.005$ kWh = 1.5 kWh per tonne, a value which has to be multiplied by a figure of 10,000 to arrive at the value required in production (approx. 18,000 kWh per tonne). Most of the excess energy is dissipated in the form of heat.

### 3.5  SUPERPURITY ALUMINIUM BY THE ZONE MELTING TECHNIQUE [11]

For scientific research into the properties of extremely pure aluminium, the metal obtained by the three-layer electrolysis process requires to be further refined.

The zone melting technique for super-refining makes use of the physics of solid–liquid equilibrium. To obtain an understanding of the principles involved it is helpful to consider the reactions which may occur in an imaginary alloy system of two components $A$ and $B$ on varying the temperature. The equilibrium diagram is a convenient and quantitative means of representing in graphical form the phases present at any specified composition and temperature as well as the changes taking place (either chemical or physical), during heating or cooling, or on changing the composition.

There is one rule that materially assists in the understanding of an alloy system: this is the Phase Rule, based on considerations of thermodynamics. If $C$ denotes the number of components and $P$ the number of phases that can exist in equilibrium, and if the presence of a vapour phase is ignored, the Phase Rule states that

$$F + P = C + 1$$

$F$ is the number of independent variables — the so-called degrees of freedom — such as temperature and concentration of components in a phase which may be changed arbitrarily without causing the disappearance of a phase or the appearance of a new one.

Fig. 23 represents the aluminium-rich portion of an equilibrium diagram

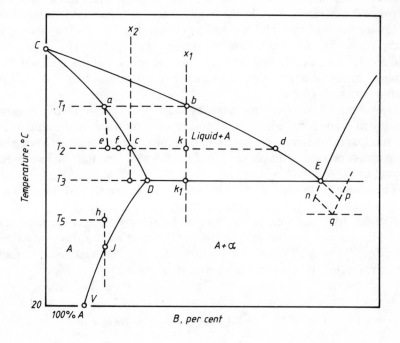

Fig. 23 — Part of an equilibrium diagram of an imaginary alloy system. (*Courtesy of Aluminium Federation.*)

of an imaginary alloy system of two components [12], aluminium and an impurity element $B$, showing the reactions which may occur on varying the temperature. $A$ is the 'terminal' solid solution, having the lattice of the pure component (in this case aluminium) with some of the $A$ atoms replaced by those of $B$. The maximum amount of $B$ capable of dissolving in $A$ increases with falling temperature while $A$ is an equilibrium with liquid, following the curve $CD$. Thereafer it decreases and at normal temperature reaches the point $V$. An alloy of composition $x_1$ will begin to freeze at the point $b$ (temperature of $t_1^0$). At this point, two phases will be present, so that the alloy will have one degree of freedom, but this is used in fixing the composition of the liquid so that the other variables — composition of solid and temperature — are automatically freed and the solid separating has the composition of the point $a$, containing a smaller percentage of the solute element $B$ than does the liquid. The liquid becomes richer in the solute, and to maintain equilibrium the temperature must fall. The composition of the liquid moves down the liquidus, reaching the composition $d$ and the temperature falling to $t_2^0$. At this moment the composition of the solid being deposited is that represented by the point $c$. In order to maintain equilibrium the composition of the solid has had to change from $a$ to $c$ following the solidus curve. This involves diffusion from liquid to solid and also within the solid itself, to ensure that the latter is homogeneous and has attained the requisite composition. At this moment the proportion of liquid to solid is given by the ratio $ck/kd$.

However, diffusion is a slow process and, if solidification is rapid, is most unlikely to occur sufficiently fast to ensure saturation and homogeneity of the solid. The primary crystals will be cored, their centres having a composition between $e$ and $c$, say $f$, and the liquid will contain an increased amount of the solute $B$.

To remove the last traces of impurities still present in refined aluminium, use is made of the physics of solid–liquid equilibrium [13].

If the liquids and solidus are lowered by the presence of an impurity solute, the equilibrium impurity content of the solid will be lower than that of the liquid, Fig. 24. The ratio of the atomic fractions in the solid and liquid phases respectively, at a given temperature, is given by

$$A_0 = \left(\frac{N_{\alpha sol}}{N_{\alpha liq}}\right) \quad \text{valid near the pure solvent end}$$

It can be shown that purification becomes easier as $A_0 \to 0$. By melting and allowing equilibrium between solid and liquid to be set up, a solid will form in which the atomic fraction $N'_{\alpha sol}$ is equal to $A_0.N_{\alpha liq}$ ($A_0$ being 0.1, for example).

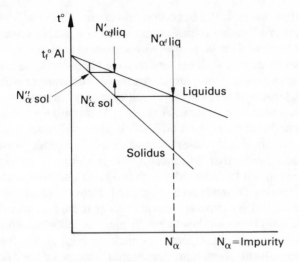

Fig. 24 — Equilibrium diagram of aluminium alloy system.

If the solid phase alone is taken and remelted to allow equilibrium to be set up between a new liquid and solid phase, the atomic fraction $N'_{\alpha\,\text{sol}}$ will be $A_0 N'_{\alpha\,\text{liq}} = A_0^2 N_{\alpha\,\text{liq}}$. Since $A_0 < 1$ it is clearly possible by repeating the sequence of operations to eliminate the impurities progressively, the ultimate contents decreasing as $A_0 \to 0$.

In practice [13], a small-diameter cylindrical bar of the refined aluminium to be super-refined is heated at one end until it melts, as indicated in Fig. 25. For this operation the bar is placed in a refractory container. Once

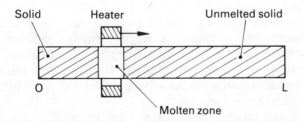

Fig. 25 — Diagram of zone melting technique of purification. (*After Van Lancker.*)

the end of the bar is molten, the molten zone is moved slowly and uniformly along the length of the bar by moving the high frequency coil used for melting, or by moving the crucible in the furnace, care being taken to produce and maintain a clearly defined and small molten zone. As the far end of the bar is reached, the operation, always in the same direction, is repeated and after a number of passes, say seven or eight, it is found that a significant proportion of the impurities with activation energies $A_0 > 1$ have been progressively moved with the molten zone towards the far end of the bar. (Those with activation energies $A_0 < 1$ move in the opposite direction.) Factors such as concentration gradient, diffusion and thermal fluctuation complicate the purification process. A series of assumptions must be made; thus it is assumed that there is no diffusion in the solid state, that perfect diffusion occurs in the liquid phase in contact with the interface (between the molten zone and solid), that no concentration gradients are set up in the liquid (this depends on the zone advance rate) and that absorption phenomena on the growing crystals can be neglected. It is also assumed that the partition coefficient $A$ is a constant, equal to $A_0$ near the pure metal (that is to say that the ratio between the atomic fractions of the solute in the solid and liquid phases at a given temperature equals the limit of $A$ as the atomic fraction of the solvent aluminium approaches unity). Furthermore, the phenomenon of 'self-slagging' due to the presence of impurities that separate out of liquid is ignored.

As the molten zone is moved along the length of the bar a condition is established such that the volume of metal solidifying in a given time is equal to that of the metal being melted. The solidifying metal will have an impurity concentration ($C_s$ related to that of the molten metal ($C_l$) which may be expressed: $C_s = kC_l$.

The value of $k$ will be different for each element and the proximity to the origin (i.e. 100 per cent Al). The presence of two or more impurity elements will modify the value of $k$ for each element. If the partition coefficient is less than unity, the solute moves away from the starting end of the bar whereas if the coefficient is greater than unity, the solute will move towards the starting end of the bar.

Thus, as the molten zone is moved along the bar by a small increment ($\delta x$) the quantity of an impurity, having a partition coefficient $< 1$, will increase in the molten zone by an amount ($C_0 - C_s)\delta x$, where $C_0$ is the concentration in the original unmelted bar. It follows from this that there will be a progressive increase in the impurity content of the molten zone and also in the solidifying metal.

The basic zone melting formulae have been described in various publications [14,15]. The equations for determining the concentrations of impurity elements at various distances $x$ from the starting end of a bar of length $L$ and cross-section 1 cm$^2$ are of the form

$$C_{s,x} \ (1st \ pass) \ = \ C_m \left[ 1 + \exp \frac{-kx}{Z}(k-1) \right]$$

where $Z$ in cm is the thickness of the molten zone and $C_m$ the concentration in the starting stock is uniform along the length of the bar. $C_s$ is the concentration in grams of solute $\alpha$ per $cm^3$ in the volume bounded by the distances $x$ and $\delta x$. After the first pass the concentration will vary along the length, as described by the above equation. Equations for the second and subsequent passes have been solved and explained by Van Lancker [15]. If sufficient passes are made, a stage will be reached at which no further refining can be achieved; this is when the quantity of impurity solidifying from the molten zone is the same as that entering the liquid from the freshly melted solid.

Using the equations it is possible to determine the position in the treated bar at which the refined metal reaches the limit for acceptable concentrations of impurities, thus defining the material which must be cut off and rejected as being too impure.

By reducing the molten zone thickness, $Z$, compared with the bar length, final distribution curves with better properties are obtained. It is claimed that with $Z = 0.1$ to $0.5\,L$ it is possible to plot the final distribution without having to go through the rather lengthy calculations.

By removing material from both ends of the bar it is possible to produce aluminium having a total impurity content of 10 ppm $= 0.001$ per cent starting from material of purity between 99.990 and 99.998 per cent, i.e. having between 100 and 20 ppm total impurity.

## REFERENCES

[1] Van Lancker, M., *Metallurgy of Aluminium Alloys*, 1967, Chapman & Hall, p. 46.
[2] Van Lancker, M., *ibid.*, p. 52.
[3] Van Lancker, M., *ibid.*, p. 51.
[4] Van Lancker, M., *ibid.*, p. 52.
[5] Mergault, P., *Bath Soc. Fr. Electr.*, Oct. 1956, p. 694.
[6] Van Lancker, M., *ibid.*, p. 50.
[7] Van Lancker, M., *ibid.*, p. 47.
[8] Van Lancker, M., *ibid.*, p. 54.
[9] Edwards, J. D., Taylor, C. S. Russell, A. S. and Maranville, L. F., Electrical conductivity of molten cryolite and potassium, sodium and lithium chlorides. *J. Electrochemical Society*, **99**, No. 12, Dec. 1952, pp. 527–535.
[10] Van Lancker, M., *ibid.*, 54.
[11] Van Lancker, M., *ibid.*, 55.
[12] *Equilibrium Diagrams of Aluminium Alloy Systems*, p. 8, The Aluminium Development Association.

[13] Van Lancker, M., *ibid.*, 56.
[14] Pfann, *Zone Melting*, 1958, Wiley.
[15] Van Lancker, M., CEN Report, 1959.
[16] Daraut, Soc. Fr. Elect. Conference, March 1962.

# 4

# Aluminium alloy systems

## 4.1 GENERAL

Aluminium of the purity obtained from the electrolytic reduction of alumina is a relatively weak material. For applications requiring greater mechanical strength it is alloyed with metals such as copper, magnesium, manganese and zinc, usually in combinations of two or more of these elements together with iron and silicon.

Large numbers of alloys have been developed to meet specific requirements [1] and the principal alloys are covered by a series of national and international standards which specify compositions and mechanical properties.

### 4.1.1 Standard specifications

The alloys covered by standards can be divided and considered in two main groups, namely those used for castings and those to be fabricated into wrought forms. For ease of reference these will be referred to by the BS 1490 LM designations for alloys for castings, and for wrought alloys by the internationally agreed four-digit system in which the first of the four digits in the designation indicates the major alloying element of alloys within the group, as follows:

| | |
|---|---|
| 1XXX | Aluminium of 99.00 per cent minimum purity |
| 2XXX | Copper |
| 3XXX | Manganese |
| 4XXX | Silicon |
| 5XXX | Magnesium |
| 6XXX | Magnesium plus silicon |
| 7XXX | Zinc |
| 8XXX | Other elements |
| 9XXX | Unused series |

A comparison of national specification designations is given in Table 14 for ingots and casting alloys and in Table 15 for wrought alloys. As both the British and the international standards for wrought materials have been changed within recent years the current and former designations are included in Table 15 in order to reduce the risk of confusion.

The chemical compositions of alloys for general engineering are given in Table 16 for castings and in Table 17 for wrought materials. Table 18 lists the forms in which wrought alloys are normally available.

The LM designations of the BS 1490 series of casting alloys are not defined or arranged in the same form of grouping as the wrought alloys. Of the twenty-three alloys listed in BS 1490, all but five (LM0, LM5, LM10, LM12 and LM31) have silicon as one of the major alloying elements.

### 4.1.2 Equilibrium diagrams
The equilibrium diagrams of aluminium alloy systems shown in figures 26 to 37 are relevant to both cast and wrought alloys.

The properties of an alloy are determined by its crystal structure which may include a number of phases differing in composition and depending on temperature. Equilibrium diagrams represent the relationships between temperature and composition of the phases in an alloy system in equilibrium, i.e. which is completely stable. Such conditions are attained by special cooling or prolonged heat treatment of alloys of exactly known purity as determined in the laboratory, but most metallic materials in industry are not in equilibrium. Nevertheless equilibrium diagrams provide guidance and a source of understanding of the reactions which take place in commercial alloys which, in addition to the main alloying elements, normally contain impurities in quantities which can have important effects on the constitution, such as introducing new phases and other modifications in the structure. Commercial alloys are seldom cast and processed under conditions which are conducive to the establishment of equilibrium, but nevertheless the diagrams provide valuable guidance to the behaviour of the alloys in practice.

In describing equilibrium diagrams, frequent reference is made to solid solutions. Van Lancker [2] writes: 'A solid solution of some element in aluminium must not be regarded as a homogeneous macroscopic entity. Certain aluminium atoms in the f.c.c. lattice can be replaced by solute atoms; certain lattice vacancies can be filled by solute atoms; and in the same way interstices can be filled'. But the f.c.c. lattice obtained in this way will reject certain solute atoms by a regular mechanism, forming a cellular structure. The solid solution (of the solvent) in aluminium thus becomes an equilibrium structure between an f.c.c. lattice containing a few impurities and a system of cellular envelopes. The envelopes are thin and do not upset the crystallographic orientation of the (111) planes in the grains; the latter remain virtually parallel. They constitute the same aluminium lattice heavily loaded with impurities, giving a cellular honeycomb structure.

In this structure there are vacancies or lacunae, groups of vacancies, dislocations and other structural defects which play just as important a part as the impurities themselves. The complex dislocation network is associated with a complex network of solute atoms (in solid solution). The decomposition of metastable solid solutions gives rise to a residual solid solution and new complex phases, and the interaction of dissolved atoms with vacancies, dislocations and the solvent leads to the formation of transitional structures known as 'zones'. The interaction between dislocations and obstacles produced by heat treatment (quenching and ageing or tempering), i.e. the zones, or by working (blocked dislocations), produces an increase in tensile strength and a reduction in ductility and plasticity.

## 4.2   1XXX GROUP: ALUMINIUM OF 99.00 PER CENT MINIMUM PURITY

The main impurities will normally be iron and silicon, with the possible presence of copper or zinc in quantities of less than 0.1 wt% and other elements in quantities of less than 0.5 per cent each, totalling less than 0.15 per cent. However, in AA1100 a small addition of copper is present as an alloying addition for the purpose of modifying corrosion characteristics.

The diagram of the binary aluminium–iron system is shown in Fig. 26. The aluminium – silicon binary alloys are dealt with in the section on 3XXX alloys (see Fig. 39).

The aluminium–iron alloys form a eutectiferous series [3] with very small solid solubility, the eutectic between the constituent designated $FeAl_3$ and aluminium lying at 1.7 per cent iron and 655°C at which temperature the aluminium-rich solid solution contains 0.052 per cent iron. The solid solubility decreases with the temperature, being 0.025 per cent at 600°C and 0.006 per cent at 500°C. However, these values can be achieved only by prolonged annealing followed by quenching, so that aluminium, even superpurity metal, will normally contain particles of iron-bearing constituents.

Silicon is also a normal impurity in aluminium, so it is necessary to study the aluminium-rich corner of the aluminium–iron–silicon ternary diagrams, Fig. 27 and Fig. 28.

Isothermal sections for temperatures below the solidus are all replicas of the phase boundaries of the solidus surface, Fig. 28. With falling temperatures the apices of the three-phase triangles move towards the aluminium corner of the model.

A ternary alloy of aluminium may contain three and possibly more

**Table 14** — Comparison of national specifications — casting alloys†

| UK BS/DTD | ISO | USA AA | USA SAE | France AFNOR | West Germ Werkstoff |
|---|---|---|---|---|---|
| LM0 | Al 99.5 | | | A5 | |
| LM2 | Al Si10 Cu2 Fe | 384.1 | 303 | A-S9 U3-Y4 | |
| LM4 | Al Si5 Cu3 | 319.2 | 326 | A-S5 U | 3.2151 |
| LM5 | Al Mg5 | 514.1 | 320 | A-G6 | 3.3561 |
| LM6 | Al Si12 | A413.2 | | A-S13 | 3.2581 |
| LM9 | Al Si12 Mg | A360.2 | 309 | A-S10 G | 3.2381 |
| LM10 | Al Mg10 | 520.0 | 324 | A-G10-Y4 | 3.3591 |
| | | | | | |
| LM12 | Al Cu10 Si12 Mg | 222.1 | 34 | | |
| LM13 | Al Si11 Mg Cu | A332.1 | 321 | A-S12 U N | |
| LM16 | Al Si5 Cu1 Mg | 355.1 | 322 | | |
| LM18 | Al Si5 | A443.1 | 35 | | |
| LM20 | Al Si12 Cu Fe | A413.1 | 305 | | 3.2583 |
| LM21 | Al Si6 Cu4 Zn | | | A-S5 U Z | 3.2151 |
| LM22 | Al Si6 Cu3 Mn | | | A-S5 U | |
| LM24 | Al Si8 Cu3 Fe | A380.1 | 306 | A-S9 U3-Y4 | 3.2161 |
| LM25 | Al Si7 Mg | 356.1 | 323 | A-S7 G | 3.2371 |
| LM26 | Al Si9 Cu3 Mg | | | | |
| LM27 | Al Si7 Cu2 | | | | |
| LM28 | Al Si19 Cu Mg Ni | | | | |
| LM29 | Al Si23 Cu Mg Ni | | | | |
| LM30 | Al Si17 Cu4 Mg | 390.0 | | | |
| LM31‡ | Al Zn5 Mg Cr | D712 | 310 | A-Z5G | |
| 4L35 | Al Cu4 Ni2 Mg2 | 242.2 | 39 | A-U4 N T | |
| 3L51 | Al Si2 Cu Ni Fe Mg | | | A-S2 U | |
| 3L52 | Al Cu2 Ni Si Fe Mg | | | | |
| 2L91 2L92 | Al Cu4 | 295.1 295.2 | 38 | A-U5 GT | 3.1841 |
| 2L99 | Al Si7 Mg | A356.2 | 336 | A-S7 G0-3 | 3.2371 |
| | | | | | |
| DTD | | | | | |
| 716B:722B | Al Si5 Mg | | | A-S4 G | 3.2341 |
| 722B:735B | | | | | |
| DTD 5008B | Al Zn5 Mg Cr | D712 | 310 | A-Z5 G | |
| DTD 5018A | Al Mg8 Zn | | | | |

†In some cases the alloys shown are not direct equivalents of the UK alloys, and the original specification should always be referred to.
‡To be added to BS 1490 in 1987.

| Germany | Italy | Sweden | Switzerland | USSR | Australia | UK |
|---|---|---|---|---|---|---|
| alloy no. | UNI no. | | | | | BS/DTD |
| | | 14 4022 | G-Al 99.5 | | | LM0 |
| | 5076 | | | | AS 307 | LM2 |
| Al Si6 Cu4 | | | | AL 6 | AP 303 | LM4 |
| -Al Mg5 | 3058 | | | AL 28 | AP 501 | LM5 |
| -Al Si12 | 4514 | 14 4261 | | | BS 401 | LM6 |
| Al Si10 Mg | 3049 | 14 4253 | G-Al Si10 Mg | AL 4 | | LM9 |
| -Al Mg10 | 3056 | | | AL 8 | AP 505 | LM10 |
| | | | | AL 27 | | |
| | 3041 | | | AL 18V | | LM12 |
| | 3050 | | | AL 30 | | LM13 |
| | 3600 | 14 4231 | | AL 5 | AP 309 | LM16 |
| | 5077 | | | | AP 403 | LM18 |
| l Si12 (Cu) | 5079 | 14 4260 | | AL 2 | AS 401 | LM20 |
| Al Si6 Cu4 | | 14 4230 | | AL 16V | | LM21 |
| | 3052 | | | AL 6 | | LM22 |
| Al Si8 Cu3 | 5075 | 14 4252 | G-Al Si8 Cu3 | | AS 313 | LM24 |
| Al Si7 Mg | 3599 | 14 4244 | G-Al Si7 Mg | AL 9 | AS 601 | LM25 |
| | 3050 | | | | | LM26 |
| | | 14 4230 | | | | LM27 |
| | 6251 | | | | | LM28 |
| | | | | AL 26 | | LM29 |
| | | | | | | LM30 |
| | 3602 | 14 4438 | | | AP 701 | LM31 |
| | 3045 | | | | | 4L35 |
| | | | | | | 3L51 |
| | 3046 | | | | | 3L52 |
| Al Cu4 Ti | 7256 | | G-Al Cu4 Ti | | | 2L91 |
| | | | | | | 2L92 |
| Al Si7 Mg | 7257 | | | | BP 601 | |
| | | | | | CP 601 | 2L99 |
| | | | | | | DTD |
| Al Si5 Mg | 3054 | | | | | 716B:722B |
| | | | | | | 722B:735B |
| | 3602 | 14 4438 | | | AP 701 | DTD 5008B |
| | | | | | | DTD 5018A |

Table 15 — Comparison of National Specifications — Wrought Alloys

| BS and International | Alloy type depicted by old ISO no. | UK former BS designation | Australia | Canada | Franc former |
|---|---|---|---|---|---|
| 1050A | AL99.5 | 1B | | | A5 |
| 1080A | Al 99.8 | 1A | | | A8 |
| 1199 | | 1 | Al199 | | A99 |
| 1200 | Al 99 | 1C | B1200 | 990 | A4 |
| 1350 | | 1E | | | A5L |
| 2011 | Al Cu6 Bi Pb | FC1 | A2011 | CB60 | A-U5 P |
| 2014A | Al Cu4 Si Mg | H15 | B2014 | CS41N | A-U4 S |
| 2017A | Al Cu4 Mg Si | | | | A-U4 |
| 2024 | Al Cu4 Mg1 | 2L97. 2L98. L109. L110 DTD5100A | | CG42 | A-U4 |
| 2031 | | H12 | | | A-U2 |
| 2117 | Al Cu2 Mg | 3L86 | | CG30 | A-U2 |
| 2218 | | 7L25 | | | |
| 2618A | | H16 | | | A-U2 ( |
| 3103 | Al Mn1 | N3 | | | |
| 3105 | | N31 | | | |
| 4043 | | N21 | B4043 | S5 | A-S. |
| 4047 | | N2 | B4047 | S12 | A-S1 |
| 5005 | Al Mg1 | N41 | A5005 | | A-G0 |
| 5056A | Al Mg5 | N6 | | | |
| 5083 | Al Mg4.5 Mn | N8 | | GM41 | A-G4.5 |
| 5154A | | N5 | C5154 | GR40 | |
| 5251 | Al Mg2 | N4 | C5152 | | A-G2 |
| 5454 | Al Mg3.6 | N51 | A5454 | GM31N | A-G2.5 |
| 5554 | | N52 | | GM31P | |
| 5556A | | N61 | | | |
| 6061 | Al Mg1 Si Cu | H20 | A6061 | GS11N | A-G S |
| 6063 | Al Mg0.5 Si | H9 | B6063 | GS10 | |
| 6082 | Al Si1 Mg Mn | H30 | | | A-S G N |
| 6101A | | 91E | | | |
| 6463 | | E6 | | | |
| 7010 | | DTD5130:5120A | | | |
| 7104 | | DTD5024:5104A:5094A | | | |
| 7020 | Al Zn4.5 Mg | H17 | | | A-Z5 |
| 7075 | Al Zn6 Mg Cu | 2L95:L160:L161:L162 | A7075 | ZG62 | A-Z5 ( |

**Table 15** — Comparison of National Specifications — Wrought Alloys

| West Germany | | Italy | Sweden | Switzerland | USSR | International No. |
|---|---|---|---|---|---|---|
| kstoff No. | DIN designation | UNI | | | | |
| 3.0255 | Al99.5 | | 14 4007 | Al 99.5 | | 1050A |
| | | | 14 4004 | | | 1080A |
| | | | | | | 1199 |
| 3.0205 | Al 99 | UNI 3567 | 14 4010 | Al 99.0 | | 1200 |
| 3.0257 | E-Al | | | E-Al 99.5 | | 1350 |
| 3.1665 | Al Cu Bi Pb | UNI 6362 | 14 4355 | Al Cu6 Bi Pb | | 2011 |
| 3.1255 | Al Cu Si Mn | UNI 3501 | 14 4338 | Al Cu4 Si Mn | AK8 | 2014A |
| 3.1325 | Al Cu Mg1 | | | | | 2017A |
| 3.1335 | Al Cu Mg2 | UNI 3538 | | Al Cu4 Mg1.5 | D16 | 2024 |
| | | | | | | 2031 |
| 3.1305 | Al Cu Mg0.5 | UNI 3577 | | | D18 | 2117 |
| | | | | | | 2218 |
| | | | | | AK4-1 | 2618A |
| 3.0515 | Al Mn | UNI 3568 | 14 4054 | Al Mn | | 3103 |
| 3.0505 | Al Mn0.5 Mg0.5 | | | | | 3105 |
| | | | | | | 4043 |
| | | | | | | 4047 |
| | | UNI 5764 | 14 4106 | Al Mg1 | | 5005 |
| 3.3555 | Al Mg5 | UNI 3576 | | | | 5056A |
| 3.3547 | Al Mg4.5 Mn | UNI 7790 | 14.4140 | Al Mg5 | | 5083 |
| | | UNI 3575 | | | AMG3 | 5154A |
| 3.3525 | Al Mg2 Mn0.3 | | | Al Mg2 | | 5251 |
| 3.3537 | Al Mg2.7 Mn | UNI 7789 | | Al Mg2.7 Mn | | 5454 |
| | | | | | | 5554 |
| | | | | | | 5556A |
| | | UNI 6170 | | | AD3 | 6061 |
| | | UNI 3569 | 14 4104 | | AD31 | 6063 |
| 3.2315 | Al Mg Si1 | UNI 3571 | 14 4212 | Al Mg Si1 Mn | | 6082 |
| | | | | Al Mg Si0.5 | | 6101A |
| | | | | | | 6463 |
| | | | | | | 7010 |
| | | | | | | 7014 |
| 3.4335 | Al Zn Mg1 | UNI 7791 | | Al Zn4.5 Mg1 | | 7020 |
| 3.4365 | Al Zn Mg Cu1.5 | UNI 3735 | Granges SM6958 | Al Zn6 Mg Cu1.5 | V95 | 7075 |

**Table 16(a)** — Chemical compositions: castings (BS 1490)

| Alloy No. | Silicon | Copper | Manganese | Magnesium | Chromium | Nickel | Zinc | Aluminium |
|---|---|---|---|---|---|---|---|---|
| | | | | Composition† (percentage of alloying elements) | | | | |
| LM0 | | | | | | | | 99.5 min |
| LM2 | 9.0–11.0 | 0.7–2.5 | | | | | | Remainder |
| LM4 | 4.0–6.0 | 2.0–4.0 | 0.2–0.6 | | | | | ,, |
| LM5 | | | 0.3–0.7 | 3.0–6.0 | | | | ,, |
| LM6‡ | 10.0–13.0 | | | | | | | ,, |
| LM9 | 10.0–13.0 | | 0.3–0.6 | 0.2–0.6 | | | | ,, |
| LM10 | | | | 9.0–11.0 | | | | ,, |
| LM12 | | 9.0–11.0 | | 0.2–0.4 | | | | ,, |
| LM13 | 10.0–12.0 | 0.7–1.5 | | 0.8–1.5 | | | | ,, |
| LM16 | 4.5–5.5 | 1.0–1.5 | | 0.4–0.6 | | | | ,, |
| LM18 | 4.5–6.0 | | | | | | | ,, |
| LM20‡ | 10.0–13.0 | | | | | | | ,, |
| LM21 | 5.0–7.0 | 3.0–5.0 | 0.2–0.6 | 0.1–0.3 | | | | ,, |
| LM22 | 4.0–6.0 | 2.8–3.8 | 0.2–0.6 | | | | | ,, |
| LM24 | 7.5–9.5 | 3.0–4.0 | | | | | | ,, |
| LM25 | 6.5–7.5 | | | 0.2–0.6 | | | | ,, |
| LM26 | 8.5–10.5 | 2.0–4.0 | | 0.5–1.5 | | | | ,, |
| LM27 | 6.0–8.0 | 1.5–2.5 | 0.2–0.6 | | | | | ,, |
| LM28 | 17–20 | 1.3–1.8 | | 0.8–1.5 | | 0.8–1.5 | | ,, |
| LM29 | 22–25 | 0.8–1.3 | | 0.8–1.3 | | 0.8–1.3 | | ,, |
| LM30 | 16–18 | 4.0–5.0 | | 0.4–0.7 | | | | ,, |
| LM31 | | | | 0.5–0.75 | 0.4–0.6 | | 4.8–5.7 | ,, |

†The composition ranges are shown as a basis for the general comparison of alloys. For details of impurity limits, reference should be made to the relevant Standard Specification.
‡The impurity levels for LM6 are lower than those for LM20.

**Table 16(b)** — Castings (aerospace)

| Bs or DTD specification | Composition† (percentage of alloying elements) | | | | | | | | |
|---|---|---|---|---|---|---|---|---|---|
| | Silicon | Iron | Copper | Manganese | Magnesium | Chromium | Nickel | Zinc | Titanium |
| 4L35 | | | 3.5–4.5 | | 1.2 –1.7 | | 1.8–2.3 | | |
| 3L51 | 1.5–2.8 | 0.8–1.4 | 0.8–2.0 | | 0.05–0.20 | | 0.8–1.7 | | |
| 3L52 | 0.6–2.0 | 0.8–1.4 | 1.3–3.0 | | 0.5 –1.7 | | 0.5–2.0 | | |
| 4L53 | | | | | 9.5 –11.0 | | | | |
| 3L78 | 4.5–5.5 | | 1.0–1.5 | | 0.4 –0.6 | | | | |
| 2L91 | | | 4.0–5.0 | | | | | | |
| 2L92 | | | | | | | | | |
| 2L99 | 6.5–7.5 | | | | 0.20–0.45 | | | | |
| L119 | | | 4.5–5.5 | 0.20–0.30 | | | 1.3–1.7 | | |
| L154 | 1.0–1.5 | | 3.8–4.5 | | | | | | 0.15–0.25 |
| 155 | | | | | | | | | 0.05–0.25 |
| DTD 716B | | | | | | | | | |
| 722B | 3.5–6.0 | | | | 0.3 –0.8 | | | | |
| 727B | | | | | | | | | |
| 735B | | | | | | | | | |
| DTD 5008B | | | | | 0.5 –0.75 | 0.4–0.6 | | 4.8–5.7 | 0.05–0.25 |

†The composition ranges are shown as a basis for the general comparison of alloys. For details of impurity limits, reference should be made to the relevant Standard Specification. Aluminium is remainder.

**Table 17** — Chemical compositions (general and supplementary engineering standards) wrought alloys

| Alloy designation | Chemical composition wt % | | | | | | | | |
|---|---|---|---|---|---|---|---|---|---|
| | Silicon | Iron | Copper | Manganese | Magnesium | Chromium | Nickel | Zinc | Others |
| 1080A | | | | | | | | | Al 99.80 min |
| 1050A | | | | | | | | | Al 99.50 min |
| 1200 | | | | | | | | | Al 99.00 min |
| 1350 | | | | | | | | | Al 99.50 min |
| 2011 | | | 5.0–6.0 | | | | | | Bi 0.20–0.6 Pb 0.20–0.6 |
| 2014A | 0.50–0.9 | | 3.9–5.0 | 0.40–1.2 | 0.20–0.8 | | | | Zr+Ti 0.20 max |
| 2024 | | | 3.8–4.9 | 0.30–0.9 | 1.2–1.8 | | | | Ti+Zr 0.20 max |
| 2031 | 0.50–1.3 | 0.6–1.2 | 1.8–2.8 | | 0.6–1.2 | | 0.6–1.4 | | |
| 2117 | | | 2.0–3.0 | | 0.2–0.5 | | | | |
| 2618A | 0.15–0.25 | 0.9–1.4 | 1.8–2.7 | | 1.2–1.8 | | 0.8–1.4 | | Zr+Ti 0.25 max |
| 3103 | | | | 0.9–1.5 | | | | | |
| 3105 | | | | 0.3–0.8 | 0.20–0.8 | | | | |
| 4043A | 4.5–6.0 | | | | | | | | |
| 4047A | 11.0–13.0 | | | | | | | | |
| 5005 | | | | | 0.50–1.1 | | | | |
| 5056A | | | | 0.10–0.6 | 4.5–5.6 | | | | Mn+Cr 0.10–0.6 |

| Alloy | Si | Cu | Mn | Mg | Cr | Zn | Remarks |
|---|---|---|---|---|---|---|---|
| 5083 | | | 0.40–1.0 | 4.0–4.9 | 0.05–0.25 | | |
| 5154A | | | 0.10–0.50 | 3.1–3.9 | | | Mn+Cr 0.10–0.50 |
| 5251 | | | 0.50–1.0 | 1.7–2.4 | | | |
| 5454 | | | 0.5–1.0 | 2.4–3.0 | 0.05–0.20 | | |
| 5554 | | | | 2.4–3.0 | 0.05–0.20 | | Be 0.0008 max for welding wire |
| 5556A | | 0.15–0.40 | 0.6–1.0 | 5.0–5.5 | 0.05–0.20 | | |
| 6061 | 0.40–0.8 | | — | 0.8–1.2 | 0.04–0.35 | | |
| 6063 | 0.20–0.6 | | | 0.45–0.9 | | | |
| 6082 | 0.7–1.3 | | 0.40–1.0 | 0.6–1.2 | | | |
| 6101A | 0.30–1.7 | | | 0.45–0.9 | | | |
| 6463 | 0.20–0.6 | | | 0.45–0.9 | | | |
| 7010 | | 1.5–2.0 | | 2.2–2.7 | | 5.7–6.7 | Zr 0.11–0.17 |
| 7020 | | | 0.05–0.50 | 1.0–1.4 | 0.10–0.35 | 4.0–5.0 | Zr 0.08–0.20  Zr+Ti 0.08–0.25 |
| 7075 | | 1.2–2.0 | | 2.1–2.9 | 0.10–0.25 | 5.1–6.4 | Ti+Zr 0.20 max |

**Table 18** — Forms in which wrought aluminium alloys are available

| Alloy designation | Plate | Sheet and strip | Drawn strip | Drawn tube | Longitudinally welded tube | Forging stock and tube | Bolt and screw stock | Rivet stock | Bar etc. | Solid conductor | Wire |
|---|---|---|---|---|---|---|---|---|---|---|---|
| 1080A | × | × | | | | | | | | | × |
| 1050A | × | × | | × | | × | | × | × | | × |
| 1200 | × | × | | × | | | | | × | × | × |
| 1350 | | | × | | | | | | × | × | × |
| 2011 | | | | | | | | | × | | × |
| 2014A | × | × | | × | | × | × | × | × | | × |
| 2024 | × | × | | | | | | | | | |
| 2031 | | | | | | × | | | | | |
| 2117 | | | | | | | | × | | | |
| 2618A | × | | | | | × | | | × | | |
| 3103 | × | × | | | | | | | | | × |
| 3105 | × | × | | | | | | | | | |
| 4043A | | | | | | | | | | | × |

| | 5056A | 5083 | 5154A | 5251 | 5454 | 5554 | 5556A | 6061 | 6063 | 6063A | 6082 | 6101A | 6463 | 7010 | 7020 | 7075 |
|---|---|---|---|---|---|---|---|---|---|---|---|---|---|---|---|---|
| × | × | | × | × | | × | × | × | × | | | | | | | |
| | | × | × | × | × | | | × | × | × | × | × | × | | × | × |
| | × | | × | | | | | | | | × | | | | | |
| | × | | | | | | | | | | × | | | | | |
| | | × | × | × | × | | | × | | | × | | | | | × |
| | | | | × | | | | | | | | | | | | |
| | | × | × | × | × | | | × | | | × | | | | | |
| × | | × | × | × | × | | | | | | × | | | | × | × |
| | | × | × | × | × | | | | | | × | | | × | × | × |

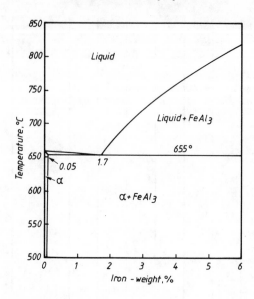

Fig. 26 — Equilibrium diagram: Aluminium–iron. (*Courtesy Aluminium Federation.*)

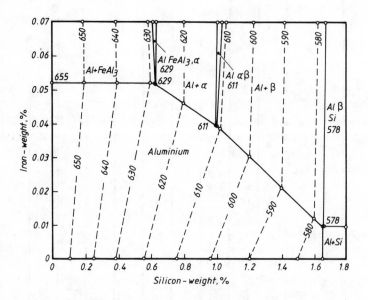

Fig. 27 — Equilibrium diagram: Aluminium–iron–silicon solidus surface. (*Courtesy Aluminium Federation.*)

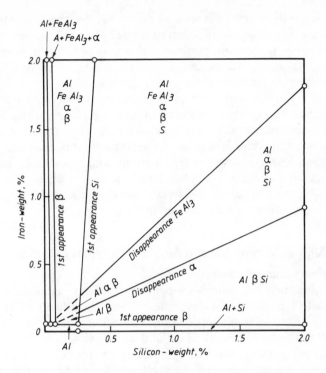

Fig. 28 — Equilibrium diagram: Aluminium–iron–silicon: metastable condition.
(*Courtesy Aluminium Federation.*)

constituents in addition to the aluminium-rich solid solution and the phases, $FeAl_3$ and silicon of the binary systems. Two of the ternary phases have fairly wide homogeneity ranges: α (FeSi) extends from 8 to 10 per cent silicon at 33 per cent iron and β (FeSi) from 13.5 to 15.5 per cent silicon at 27 per cent iron: δ (FeSi) has a small homogeneity range round the composition 25 per cent silicon–25.5 per cent iron.

Undercooling can result in a shift of the primary field boundaries. Rapid solidification, as experienced in commercial practice, prevents the aluminium phase from taking up iron or silicon to the saturation point, with the result that the liquid phase becomes enriched in these elements and deposits them in the form of eutectics. Silicon therefore appears in the microsections if the amount present exceeds about 0.25 per cent, and $FeAl_3$ appears at about 0.02 per cent iron (Fig. 28). In commercially pure aluminium, where silicon is invariably present, one or both of the ternary compounds α (FeSi) and β (FeSi) may occur.

By far the largest tonnage of aluminium is supplied to specifications

having composition limits covered by the 1XXX designations. Many commodities are fabricated from selected grades of metal supplied from the
reduction plants without any additions other than possibly small amounts of
such elements as boron for the improvement of electrical conductivity values
or boron/titanium for refining the as-cast grain size.

For some applications, such as electrical conductor wire, additions of
about 1 per cent iron may be made to increase mechanical strength without
seriously impairing conductivity. High-iron alloys of this type also find
application as thin sheet for containers and as foil for packaging.

Alloys of this series have low tensile strength and good elongation values
in the annealed condition. Higher tensile strength values with reduced
elongation values can be obtained by cold working, e.g. cold rolling for sheet
or reduction by drawing for tube, rod or wire.

### 4.3   2XXX GROUP: ALUMINIUM ALLOYS WITH COPPER

#### 4.3.1   Binary aluminium–copper alloys

In aluminium-rich aluminium–copper binary alloys, aluminium forms a
eutectic with the constituent designated $\theta$. This constituent melts at 591°C
and has no fixed stoichiometric composition but varies slightly. At this
temperature the homogeneity range of $\theta$ is from 46.4 to 47.8 wt% copper.
The ideal composition for $CuAl_2$ is 46.1 wt% copper, just outside the
homogeneity range; however, $\theta$ is often referred to by this chemical
formula. There is appreciable solid solubility of it at the aluminium end, the
solid solution containing 5.7 wt% copper at the eutectic temperature under
conditions of equilibrium, falling to 0.45 wt% at 300°C and 0.1–0.2 wt% at
250°C: Fig. 29.

Van Lancker [2] states that on casting a typical 4 per cent copper alloy, a
solid solution of cellular structure is obtained in which copper is not
distributed randomly. On prolonged heating at temperatures above the
solid solubility curve but below the solidus, the copper in $\alpha$ aluminium is
redistributed in a complex manner while lattice vacancies are concentrated
to a notable extent. Quenching or extra high cooling rates from a temperature of 540°C freeze the copper and vacancies *in situ*.

On tempering the quenched alloy at 150°C the solid solution decomposes
into another solid solution plus excess copper as clusters which form the
nuclei for the formation of Type 1 Guinier-Preston zones [4] subsequently
giving rise to Guinier-Preston compounds as platelets 5 Å thick and 100 Å
across. These are designated $\theta''$ and have a tetragonal lattice with
$a_0=b_0=4.04$ Å and $c_0=7.09$ Å.

The platelets lie parallel to the {100} planes in the $\alpha$ aluminium, an
arrangement dictated by the high value of $c_0$. the adjacent layers of $\alpha$
aluminium surrounding each expanded $\theta''$ will be deformed (in the $c_0$

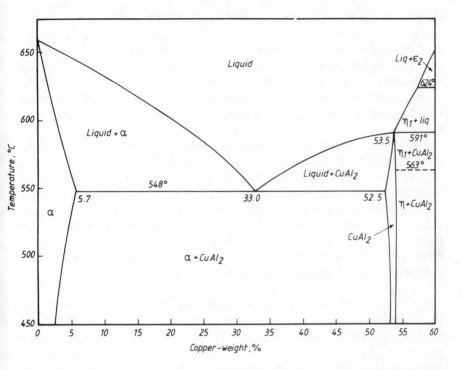

Fig. 29 — Equilibrium diagram: Aluminium–copper. (*Courtesy Aluminium Federation.*)

direction of the platelet) by elastic coherency stresses resulting in structural hardening by internal stresses, reaching a maximum when the entire $\alpha$ structure is in a state of stress.

On raising the tempering temperature to 200°C, tetragonal $\theta'$ platelets are formed and $c_0$ decreases to 5.08 Å. These platelets are about 200 Å thick and are parallel to the {100} faces of the $\alpha$ aluminium, forming three orthogonal systems.

The $\theta'$ phase then develops further towards tetragonal $\theta$ which is responsible for softening the alloy by eliminating the internal stresses since the $\theta$ structure is very different from the f.c.c. structure of the aluminium and the $\theta$ crystals consequently behave as inclusions to grow into spheroidal form. By losing coherency with the $\alpha$ lattice, the $\theta$ crystals relieve the internal stresses and the high mechanical properties disappear. Since $\theta$ is non-coherent, the dislocations by-pass the crystals rather than pass through them.

At low cooling rates from high temperatures, vacancies will only partially survive to low temperatures; consequently copper will diffuse more slowly and the hardening process will be retarded and reach a lower peak value.

Commercial aluminium–copper alloys contain iron and silicon as the principal impurities; both have a marked structural effect. Depending upon the amount present, iron will occur as $FeAl_3$, as one or both of the aluminium–iron–silicon ternary compounds $\alpha$ (FeSi) or $\beta$ (FeSi), or as the ternary aluminium–iron-copper compound $\beta$ (FeCu). Silicon will occur as $\alpha$ (FeSi), $\beta$ (FeSi) or in elemental form.

Iron which forms highly insoluble constituents is of great value in compositions for elevated temperature applications.

Binary aluminium–copper alloys are not used commercially other than as master alloys for the making of more complex alloys. Wrought alloys of the aluminium–copper group have additions of elements such as magnesium, manganese, silicon and nickel together with smaller additions of titanium, chromium or zirconium. These alloys are widely used for structural applications in the aircraft industry and in general engineering when light weight combined with strength is required.

### 4.3.2   Ternary aluminium–copper–magnesium alloys

In the aluminium–copper–magnesium alloys the constituents which can exist in equilibrium with aluminium-rich solid solution are the two binary phases $Mg_2Al_3$ — see Fig. 29 and Fig. 30 — and $CuAl_2$, and two ternary phases designated S and T. The two latter have primary phase fields covering a wide range, but elsewhere are formed as the result of a peritectic reaction. The phase S has a small homogeneity range and is based on the formula $CuMgAl_2$. The phase T has a wide homogeneity range and has a composition approximating to $CuMgAl_6$.

Aluminium and the S phase form what is practically a quasi-binary system. The boundary between these two primary phase fields rises to a maximum on crossing the quasi-binary line at 518°C, 24.5 per cent copper, 10.1 per cent magnesium. On the copper-rich side of the quasi-binary line there is a ternary eutectic of Al, S phase and $CuAl_2$ freezing at 507°C and containing 33 per cent copper, 6 per cent magnesium. Figs. 31(a) and 31(b) show sections of the ternary diagram.

Over a fairly wide range of compositions the only constituent separating from solid solution is the S phase and this is therefore the age-hardening agent.

Along the quasi-binary section running from the aluminium corner of the liquids diagram, Fig. 31(b), to the maximum on the Al–S boundary, S is the only metastable constituent to appear in alloys rapidly solidified. It can be identified in alloys containing as little as 0.5 per cent magnesium, 1.2 per cent copper: it dissolves, however, on solution heat treatment. With copper, as little as 0.25 per cent in excess of that corresponding to its quasi-binary

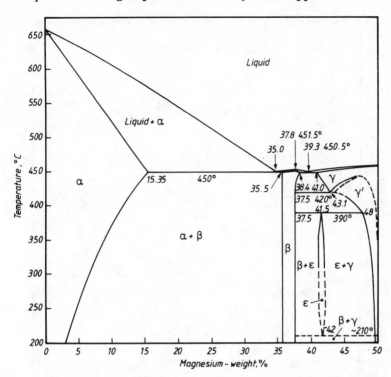

Fig. 30 — Equilibrium diagram: Aluminium–magnesium. (*Courtesy Aluminium Federation.*)

line, freezing continues until the ternary eutectic point is reached, and $CuAl_2$ is visible together with S, in the microsections. Here again, equilibrium can be attained rapidly by heat treatment. With magnesium in excess of that required for the quasi-binary line, to the amount of 1 per cent or more, freezing follows the branch of the Al–S boundary passing through the peritectic point at 467 °C where the S phase is converted more less completely into the T phase and terminating at the ternary eutectic point at 451°C where $Mg_2Al_3$ appears as a constituent (see Fig. 31(c)). It is therefore common to find residual crystals of S with a reaction rim of T in addition to $Mg_2Al_3$ in the magnesium-rich alloys of the system. All these metastable conditions, Fig. 32, disappear on heat treatment, and this occurs more rapidly in fabricated alloys where the duplex S–T crystals have been broken up by mechanical working.

### 4.3.3   Effect of other alloying additions and of minor impurities on aluminium–copper alloys
#### 4.3.3.1   *Manganese*
When present as an impurity only, this is taken into solid solution in

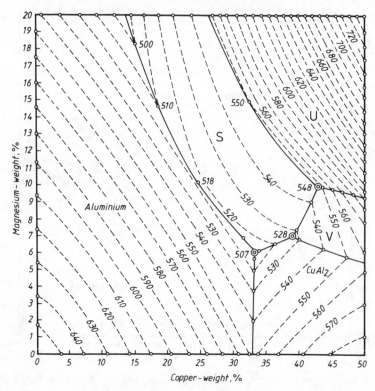

Fig. 31a — Equilibrium diagram: Sections of the aluminium–copper–magnesium
ternary alloy system. (*Courtesy Aluminium Federation.*)

aluminium (see Fig. 36); the secondary constituent is script-like $\alpha$ (FeSi).
This reacts preferentially to form $\beta$ (CuFe) liberating silicon which forms
$Mg_2Si$ and later still, the quaternary phase $\lambda$ based on the formula
$Al_5Cu_2Mg_8Si_6$ [5]. In alloys where the silicon content is of the same order or
exceeds the iron content, the $FeAl_3$ phase does not occur.

When manganese is present to the extent of 0.25 per cent or more, the
secondary constituent is $MnAl_6$ separating at a temperature only a few
degrees below that at which aluminium begins to crystallise. The phase
$MnAl_6$ is capable of taking a considerable quantity of iron into solid
solution, and unless the iron content is restricted, there is a risk of the
primary field being entered, with the consequent formation of relatively
large crystals. At a lower temperature, $MnAl_6$ can react peritectically in two
ways, with copper to form $\alpha$ (Cu–Mn) or with silicon to form $\alpha$ (Mn–Si); the
latter appears to be the more stable constituent and forms preferentially. It
forms an unbroken series of solid solutions with the corresponding body of

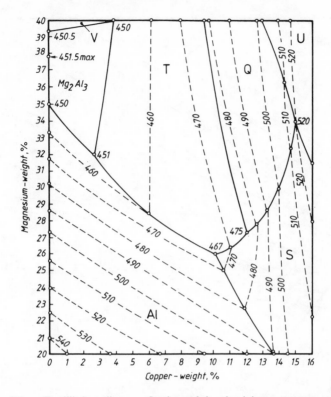

Fig. 31b — Equilibrium diagram: Sections of the aluminium–copper–magnesium ternary alloy system. (*Courtesy Aluminium Federation.*)

the Al–Fe–Si ternary system $\alpha$ (FeSi) and it crystallizes in a characteristic script formation designated $\alpha$ (Fe, Mn–Si). The peritectic reaction rarely goes to completion, so that residual $MnAl_6$ is a frequent constituent of the microsections, and sufficient silicon remains in the liquid phase to form $Mg_2Si$. The reaction between Mg and Si to form $Mg_2Si$ is a balanced one and requires an excess of one or more reactants for its completion. It is not uncommon to find particles of free silicon in the microstructure, particularly if the magnesium content of the alloy is low.

### 4.3.3.2  Nickel

The aluminium-rich binary alloys of the aluminium–nickel series, Fig. 33, form a eutectic at 640°C and 6.4 per cent nickel. This is formed between the solid solution of nickel in aluminium which is low (about 0.05 per cent at 640°C) and the the compound $NiAl_3$ which forms at 854°C by peritectic reaction. Tempering after quenching from high temperature does not

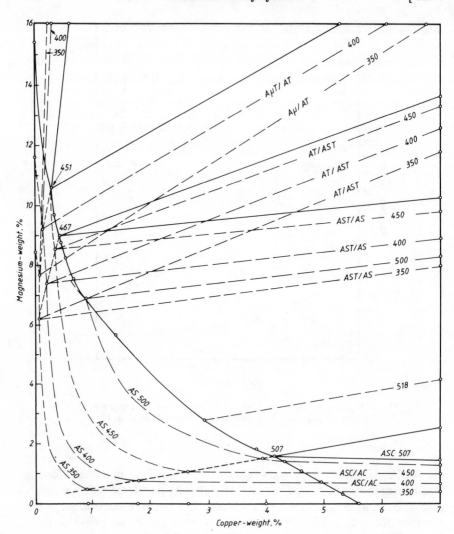

Fig. 31c — Equilibrium diagram: Sections of the aluminium–copper–magnesium ternary alloy system. (*Courtesy Aluminium Federation.*)

produce sufficient hardening of binary alloys to be of practical use but the more complex alloys of the aluminium–copper–magnesium–iron–nickel– silicon type are used for forgings, having good creep resistant properties at elevated temperatures. Nickel additions also have the effect of reducing hot shortness in castings and of reducing the coefficient of thermal expansion.

Nickel was an essential element in Y alloy (4 per cent Cu, 2 per cent Ni,

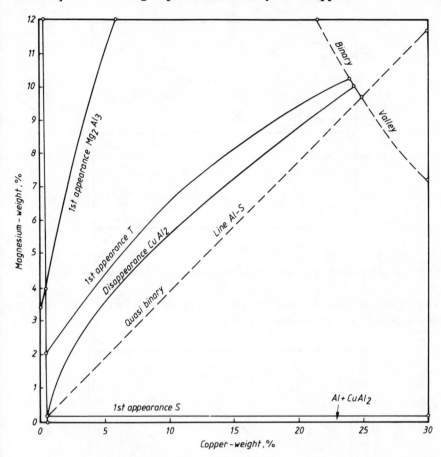

Fig. 32 — Equilibrium diagram: Aluminium–copper–magnesium: metastable.
(*Courtesy Aluminium Federation.*)

$1\frac{1}{2}$ per cent Mg, balance Al) developed during the 1914–18 War by the NPL under Dr Rosenhain as an alloy which would retain its tensile strength at comparatively high temperatures for use in castings for aero engines. It was further developed in the mid-1920s as an alloy for forged pistons.

### 4.3.3.3 Titanium
As shown in Fig. 34 the solid solution of titanium in aluminium is 1.15 per cent at the peritectic temperature of 665°C, the reactants being an intermetallic phase $TiAl_3$ and liquid containing 0.15 per cent Ti. The solid solubility

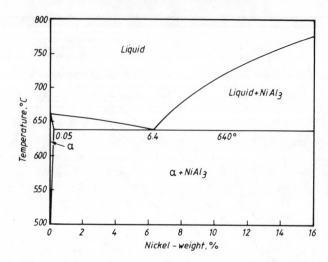

Fig. 33 — Equilibrium diagram: Aluminium–nickel. (*Courtesy Aluminium Federation.*)

of titanium in solid aluminium decreases rapidly with falling temperature, being as little as 0.24 per cent at 510°C.

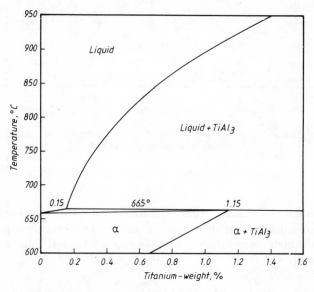

Fig. 34 — Equilibrium diagram: Aluminium–titanium. (*Courtesy Aluminium Federation.*)

### 4.3.3.4 Zirconium
This produces an equilibrium diagram of a similar type to that of Al–Ti. The solid solubility of zirconium in aluminium is only 0.28 per cent at the

peritectic temperature of 660.5°C where reaction occurs between melt with 0.11 per cent Zr and ZrAl$_3$. The solid solution is about 0.05 per cent Zr at 500°C.

Zirconium does not refine the as-cast grain structure but the recrystallized structure of wrought alloys after working and thermal treatment is finer than that of similar material but without the zirconium addition.

### 4.3.3.5  Chromium
Like titanium in aluminium, chromium undergoes a peritectic reaction at 661°C between liquid with 0.4 per cent Cr and solid CrAl$_7$. The solid solubility is 0.77 per cent at 661°C and 0.15 per cent at 300°C — see Fig. 35.

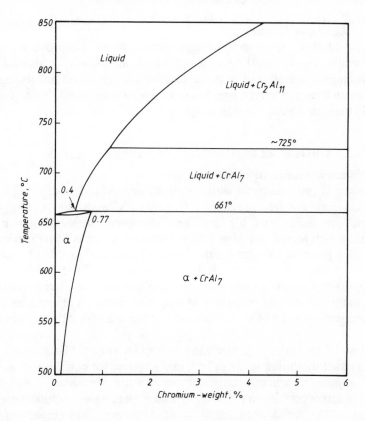

Fig. 35 — Equilibrium diagram: Aluminium–chromium. (*Courtesy Aluminium Federation.*)

In alloys having chromium contents between 0.4 and 0.77 per cent, the peritectic reaction between CrAl$_7$ and liquid at 661°C to give aluminium-rich solid solution requires appreciable time for its completion. Dependent upon cooling conditions from the molten state the constituent $CrAl_7$ may appear in the microsection at percentages of chromium well below 0.4.

Iron is taken into solution in $CrAl_7$ and any excess separates as $FeAl_3$ containing a little chromium. Silicon reacts to form a ternary compound designated $\alpha$ (CrSi). Manganese is taken into solid solution in $CrAl_7$ but over certain ranges of composition a constituent designated G occurs. This is a metastable constituent of the Mn–Al binary system, stabilised by dissolved chromium. Magnesium is not appreciably soluble in $CrAl_7$ and a ternary constituent, designated E, is found in all alloys in which the magnesium content exceeds about 1 per cent and the chromium content 0.5 per cent. Zinc reduces the solubility of chromium in aluminium.

### 4.3.3.6  *Lead and bismuth*
These are additions made to aluminium–copper alloys to improve machinability. Small amounts (0.01 per cent) of lead as an impurity in aluminium–copper–magnesium alloys can cause extensive cracking on hot rolling [6] owing to melting of the lead which is insoluble in aluminium. It also causes hot shortness in Al–Zn–Cu–Mg alloys of the 7075 type [20].


## 4.4  3XXX GROUP: ALUMINIUM–MANGANESE ALLOYS
### 4.4.1  Binary aluminium–manganese alloys
Aluminium forms a eutectic with the constituent $MnAl_6$ containing 2 per cent manganese and freezing at 658.5°C, only 1.5°C below the freezing point of pure aluminium — see Fig. 36. There is appreciable solid solubility in aluminium-rich alloys, reaching 1.82 per cent at the eutectic temperature. With 0.015 per cent iron the solid solubility is reduced to 1.38 per cent at 650°C [4]. As the temperature falls the manganese concentration falls to 0.95 per cent at 600°C and 0.35 per cent at 500°C. The manganese is rejected from the aluminium lattice as secondary $MnAl_6$. This constituent contains 25.34 wt% manganese and has an orthorhombic structure with 28 atoms per unit cell.

Alloys in this system are rarely in equilibrium as cast. Manganese is not taken into solid solution to the full amount required for equilibrium, so that $MnAl_6$ appears as a micro-constituent at very low percentages of manganese. Undercooling is very prevalent and supercooling produces strong supersaturation which may result in an apparent displacement of the eutectic to 3–4 per cent and the $MnAl_6$ crystallises in a high degree of dispersion. The peritectic reaction at 710°C involving the conversion of $MnAl_4$ to $MnAl_6$ is extremely subject to undercooling and may on occasion be suppressed completely so that $MnAl_4$ cores may occur at percentages of manganese rather lower than 4.1 or they may not appear until the manga-

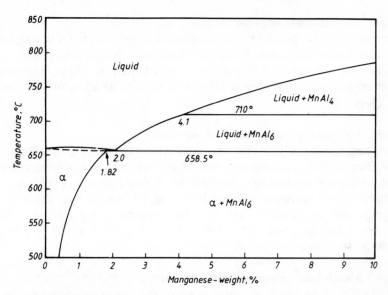

Fig. 36 — Equilibrium diagram: Aluminium–manganese. (*Courtesy Aluminium Federation.*)

nese content is of the order of 6 per cent. At low percentages of manganese, a metastable constituent G may occur, whiter in colour than $MnAl_6$ but the mode of formation is not known.

### 4.4.2   Effect of impurities
The impurities commonly present in aluminium–manganese alloys are iron, silicon and copper. However, in some alloys of this group, copper contents ranging from 0.20 to 0.30 per cent maximum are specified, and this element has a significant effect on corrosion characteristics. Iron is taken into solid solution in $MnAl_6$ and manganese in $FeAl_3$. As mentioned earlier, iron has the effect of reducing the solid solubility of manganese in aluminium–manganese alloys and additions of 0.6 per cent iron may be made to 1.2 per cent manganese for the purpose of inhibiting grain growth on annealing wrought material.

### 4.4.3   Aluminium–manganese–silicon
In the system aluminium–manganese–silicon there is a ternary constituent designated $\alpha$ (MnSi) which forms a eutectic with aluminium in which $\alpha$ (MnSi) crystallizes in script formation; the eutectic valley terminates in a ternary eutectic of aluminium, $\alpha$ and silicon at 573°C. The $\alpha$ phase is formed peritectically from $MnAl_6$ so that, in the presence of silicon, crystals of $MnAl_6$ tend to be surrounded by reaction rims of $\alpha$. A similar peritectic reaction occurs in aluminium–manganese alloys containing copper.

### 4.4.4  Aluminium–manganese–magnesium
Some of the alloys of the 3XXX group have magnesium contents ranging up to 1.5 per cent maximum. The purpose of this addition is to improve the mechanical properties of this series of work hardening alloys. The aluminium–magnesium alloys are dealt with in greater detail in the section on 5XXX group.

### 4.4.5  Aluminium–manganese (~4 per cent)
Manganese ($\simeq$ 4 per cent) forms a most important alloying element in the 'Al–Tran†' alloys [2], supercooling or liquid quenching producing a strong supersaturation of manganese in aluminium. These alloys are claimed to be refractory, anti-corrosive and suitable for high temperatures. Pressure die castings can be anodised and the alloys can be transformed into wire extrusions and forgings. The Al–Tran alloys are not covered by international specifications.

### 4.4.6  Aluminium–manganese alloy products
The 3XXX alloys are of medium strength and are normally fabricated as sheet or plate for applications such as building sheet or panels for roofing and sidings, domestic utensils, etc.

### 4.5  4XXX GROUP: ALUMINIUM–SILICON ALLOYS
#### 4.5.1  Binary aluminium–silicon alloys
Aluminium and silicon form a simple eutectiferous series with some solid solubility at both ends — Fig. 37. The eutectic of aluminium and silicon contains 11.7 per cent silicon and freezes at 577°C. The aluminium-rich constituent, under conditions of equilibrium, contains 1.65 per cent silicon at this temperature and the silicon-rich constituent 0.5 per cent aluminium. The solid solubility of silicon in aluminium decreases to 1.3 per cent at 550°C, 0.8 per cent at 500°C, 0.29 per cent at 400°C and 0.05–0.008 per cent at 250°C.

Two forms of silicon can exist in the alloys:

 (i) that resulting from precipitation from $\alpha$ solid solution, and
(ii) that produced by direct solidification from the eutectic melt.

The two are crystallographically equivalent but differ in form and distribution. There are no intermetallic compounds between aluminium and silicon.

As ordinarily cast, aluminium–silicon alloys are not in structural equili-

---

† Patent, registered alloys of aluminium and transition metals.

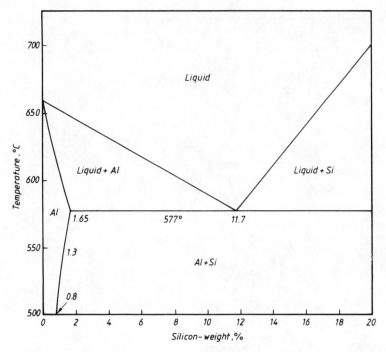

Fig. 37 — Equilibrium diagram: Aluminium–silicon. (*Courtesy Aluminium Federation.*)

brium: the aluminium-rich solid solution is cored and free silicon may be detected in microsections when the silicon content is as low as 0.25 per cent.

### 4.5.2 Effect of other elements

Of the other elements which may be present, the most important from the point of view of structure and mechanical properties are sodium and iron. The former is intentionally added to the melt just prior to casting to produce the reaction known as 'modification'. The sodium causes the crystallization of the silicon to be suppressed, with the result that the eutectic is shifted to some 14 per cent of silicon and lowered in temperature, whilst the silicon constituent, instead of freezing in large, thin fragile plates, becomes highly dispersed, giving greatly improved strength, as shown in Figs 38(a) and (b). If too great a quantity of sodium is added, the silicon particles tend to become coarser, and the outer portions of each eutectic grain tend to be denuded of silicon, causing deterioration in the mechanical properties.

### 4.5.3 Effect of iron

Iron is virtually insoluble in these alloys and occurs as the ternary compound β (FeSi). If the iron is less than 0.6 per cent, the compound occurs as small

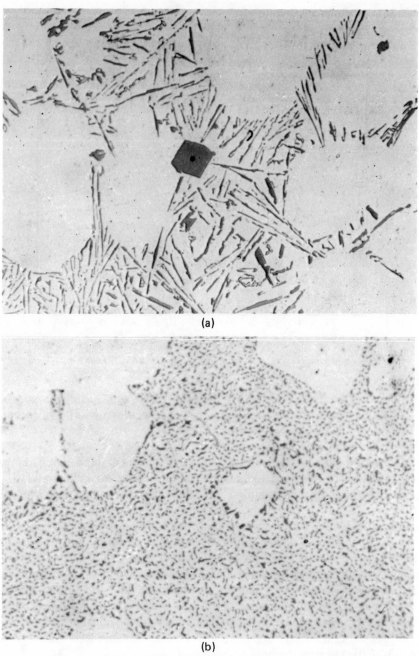

(a)

(b)

Fig. 38 — (a) Cast aluminium–12% silicon alloy before modification, showing large
plate-like silicon crystals. × 600. (b) Cast aluminium–12% silicon alloy after sodium
addition, giving the silicon eutectic. × 600.

needles and plates in the eutectic; at a rather higher value it occurs in massive form causing brittleness and so-called 'coarse crystalline' fracture, leading to a very marked deterioration in the mechanical properties. Manganese present as an impurity, or as an alloying element, combines with the silicon and iron to form a constituent, isomorphous with $\alpha$ (FeSi), which is tough rather than brittle and therefore tends to reduce the deleterious effect of high iron.

### 4.5.4 Aluminium–silicon foundry alloys
Aluminium–silicon alloys are of particular value to the foundry industry because of their high fluidity. Silicon reduces hot shortness on freezing and reduces the coefficient of thermal expansion, whilst copper and nickel improve elevated temperature properties.

The main applications for aluminium–silicon alloys in wrought forms are as welding rod and low melting point cladding for brazing quality sheet, and also for architectural applications — the anodised sheet has an attractive dark-grey colour.

The alloy 4032, having 11 per cent silicon and 1 per cent each of iron, copper, magnesium and nickel, is an alloy having a low coefficient of thermal expansion. It responds to solution heat treatment followed by precipitation treatment at elevated temperatures and is used for forgings such as cylinder barrels for radial aircraft and other i.c. engines.

## 4.6  5XXX GROUP: ALUMINIUM–MAGNESIUM ALLOYS
### 4.6.1  Binary aluminium–magnesium alloys
The aluminium-rich binary aluminium–magnesium alloys form a eutectiferous series with appreciable solid solubility at the aluminium end (Fig. 30 [3]). The eutectic of which the second constituent is that designated $\beta$ or $Mg_2Al_3$ lies at 35 per cent Mg and 450°C. The aluminium-rich solid solution contains 15.35 per cent magnesium at the eutectic temperature, falling to 11.8 per cent at 400°C, 6 per cent at 300°C, 4 per cent at 200°C and about 2 per cent at 100°C [7]. This fall in solubility is accompanied by the rejection of $Mg_2Al_3$ from the $\alpha$ solid solution. For precipitation of $\beta$, the alloy must be treated at a temperature of 200–300°C and the time of treatment is dependent upon the exact temperature level. Precipitation occurs preferentially on the {100} planes followed by the {120} planes. The precipitation may be continuous or discontinuous, depending upon the tempering temperature used; 'discontinuous' precipitation is accompanied by the formation of a new solid solution, whereas 'continuous' precipitation generates a 'Wiedmanstatten structure (needles of constituent precipitation in three cyrstallographic directions), the scale of which decreases as the tempering temperature is raised. The overall hardening effect is slight. The $\beta$ phase plays an

important role in corrosion phenomena (particularly stress corrosion of aged alloys).

For optimum corrosion resistance, the β phase appears as discrete particles, as shown in Fig. 39 ((a) and (b)) whilst after ageing (sensitising) treatment, the β phase is found at the grain boundaries, as shown in Fig. 39 ((c) and (d)). In certain cases, corrosion appears to originate in submicroscopic layers (preceding the visible precipitation of $Mg_2Al_3$) enriched or impoverished with respect to solute elements [1].

Equilibrium is seldom attained during solidification. It is not uncommon for those portions of the liquid that are last to solidify to become so enriched with magnesium that the eutectic composition is reached and β ($Mg_2Al_3$) may appear as a constituent in commercial alloys having a magnesium content of the order of 3–4 per cent when solidified rapidly.

The β phase has a complex f.c.c. lattice: this constituent is responsible for the heat treatability of the aluminium–10 per cent magnesium casting alloy. The solidification range of this alloy is 90°C (liquidus 610°C, solidus 520°C). Below 520°C a solid solution of magnesium in aluminium is formed: from 360°C downwards this precipitates the β phase in increasing amounts, and this β phase can be dissolved only by heating to about 425 °C to give the optimum mechanical properties and resistance to corrosion. This alloy requires special skill and experience in the foundry.

### 4.6.2 Precipitation in aluminium–magnesium–silicon alloys
In the alloy systems, aluminium–copper and aluminium–magnesium–silicon, precipitation is considered as an indication of lattice hardening but in the case of aluminium–magnesium alloys 'age-softening' takes place because of the large particle size instead of fine dispersion of the precipitate.

### 4.6.3 The effect of chromium and manganese additions
Chromium, manganese or a combination of these two elements are added to aluminium alloys as stress corrosion controlling elements, both elements being effective in reducing grain boundary potential. They also increase strength at the expense of a certain amount of ductility and formability.

Hume-Rothery, Raynor and their colleagues [8] in studies of alloys of commercial interest, 0 to 6 per cent magnesium, 0 to 3 per cent manganese, have shown that at the liquidus the $MnAl_6$ primary field is entered at 1.04 per cent manganese at 5 per cent magnesium, compared with 1.97 per cent manganese at 0 per cent magnesium. The solidus diagram over the same range shows that at 5 per cent magnesium, the field, over which solidification ends with simultaneous separation of $MnAl_6$ and Al, is entered at about 0.2 per cent manganese at 580°C.

An isothermal section, showing the solid alloys at 400°C, is reproduced in Fig. 40.

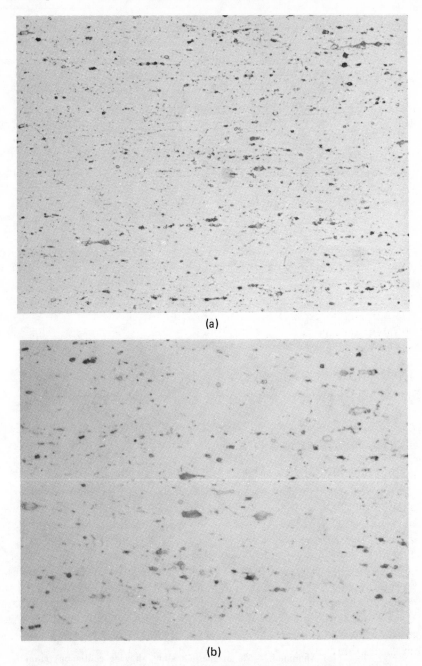

(a)

(b)

Fig. 39a — (a) Aluminium/4–5% magnesium alloy, showing satisfactory structure. × 500.  (b) Aluminium/4–5% magnesium, showing satisfactory structure. × 1000. Both etched in orthophosphoric acid.

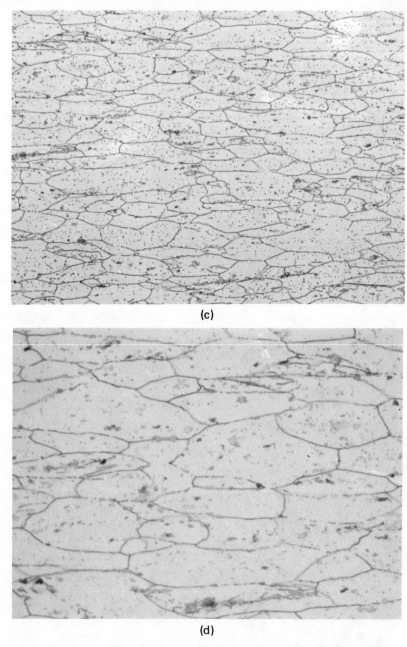

(c)

(d)

Fig. 39b — (c) Aluminium/4–5% magnesium alloy, showing continuous grain boundary network of precipitates, after sensitising treatment. × 500. (d) Aluminium/4–5% magnesium alloy, showing continuous grain boundary network of precipitates, after sensitising treatment. × 1000. Etched as above.

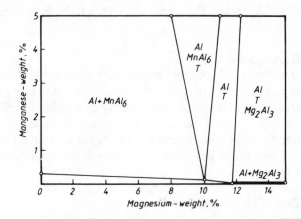

Fig. 40 — Aluminium–magnesium–manganese isothermal at 400°C.

### 4.6.4   The effect of impurities

The two most common impurities in aluminium–magnesium alloys are silicon and iron. The former combines with magnesium to form the binary compound $Mg_2Si$ and this can often be seen in microsections as small particles, grey or blue in colour. Silicon combines preferentially in this manner, so that little is available for the formation of other silicides. Consequently iron, when present as an impurity, tends to occur as $FeAl_3$ in preference to the ternary compounds $\alpha$ (FeSi) and $\beta$ (FeSi).

### 4.6.5   The effect of sodium

Sodium is an impurity which must be avoided in wrought aluminium–magnesium alloys. It can be introduced by the use of sodium–bearing metal from the electrolytic reduction process, or by the use of unsuitable sodium-bearing cover fluxes during melting. In hot worked aluminium–magnesium alloys, sodium produces embrittlement by forming isolated pores and cracks. Plasticwork at elevated temperatures generates excess lattice vacancies and a disordered grain boundary structure where sodium concentrates in free form, particularly in alloys with more than 2 per cent magnesium. The effect is major at 4 per cent magnesium and can lead to excessive cracking both during rolling and of the hot rolled product. The maximum tolerable amount of sodium is 0.001 per cent.

Ramsley and Talbot [9] have shown that in the absence of magnesium, the sodium combines with silicon to form a ternary compound NaAlSi which has a low solid solubility and as such is harmless; in the presence of magnesium, however, $MgSi_2$ is formed preferentially. Thus, it can be supposed that it is 'free' magnesium in solid solution in aluminiun that liberates the sodium from the compound in which it was previously harmless.

### 4.6.6 The effect of beryllium
Additions of 0.003 to 0.010 per cent beryllium are often made to aluminium alloys used for castings, both to reduce the oxidation and to bring down the hydrogen content of the cast metal. For welding electrodes and filler rod, a maximum of 0.0008 per cent beryllium is specified because of the health hazard associated with beryllium fumes. In wrought alloys the purpose for the addition of beryllium is to neutralise the deleterious effect of the presence of sodium.

### 4.6.7 The effect of bismuth and antimony
The harmful effect of sodium in aluminium–magnesium alloys can be removed by a ten-fold excess of bismuth [10], and an addition of antimony has a similar effect [11]. Neither sodium, bismuth nor antimony is miscible with aluminium to any measurable extent, and both bismuth and antimony form high melting point stoichiometric compounds with sodium. It is therefore probable that in the presence of sufficient of either of the two former elements all the sodium is converted from the free to the combined state.

## 4.7   6XXX GROUP: ALUMINIUM–MAGNESIUM–SILICON ALLOYS

### 4.7.1   General
The equilibrium diagrams of the binary aluminium–magnesium (Fig. 30) and aluminium–silicon systems are shown in Fig. 37 and of the ternary aluminium–magnesium–silicon liquids surface in Fig. 41, the solidus in Fig. 42 and the limits of solubility in Fig. 43.

Aluminium and the binary constituent $Mg_2Si$ form a quasi-binary system dividing the ternary system into two parts. In the quasi-binary system the two constituents form a eutectic at 595°C containing 8.15 per cent Si and there is appreciable solid solubility at the aluminium end reaching 0.85 per cent Mg, 1.10 per cent Si at the eutectic temperature; both partial ternary systems are eutectiferous. Aluminium, silicon and $Mg_2Si$ form a ternary eutectic containing 4.97 per cent magnesium, 12.95 per cent silicon and freezing at 555°C while aluminium, $Mg_2Si$ and $Mg_2Al_3$ form one at 33.2 per cent magnesium, 0.37 per cent silicon, freezing at 451°C. The primary $Mg_2Si$ field is ridge-shaped.

The single-phase field, over which aluminium is the only constituent to solidify, terminates at 15.3 per cent magnesium and 0.1 per cent silicon at 415°C. Full co-ordinates of the invariant points are given in Table 19.

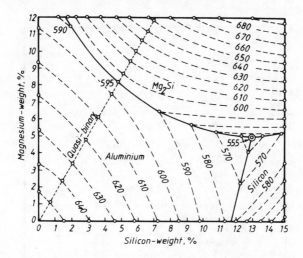

Fig. 41 — Aluminium–magnesium–silicon liquidus.

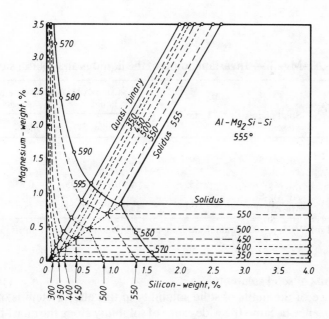

Fig. 42 — Aluminium–magnesium–silicon solidus.

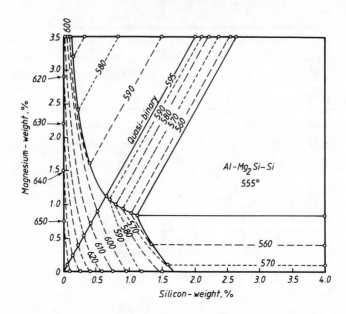

Fig. 43 — Aluminium–magnesium–silicon: limits of solubility.

**Table 19** — Al–Mg–Si — Invariant points of the liquidus and solidus surfaces

| Description | Solid phases present | Temp.°C | Composition of participating liquid | | Al–rich solid | |
|---|---|---|---|---|---|---|
| | | | Mg% | Si% | Mg% | Si% |
| Quasi-binary eutectic | Al, Mg₂Si | 595 | 8.15 | 4.75 | 1.13 | 0.67 |
| Ternary eutectic | Al, Mg₂Si, Si | 555 | 4.97 | 12.95 | 0.85 | 1.10 |
| Ternary eutectic | Al, Mg₂Si, Mg₂Al₃ | 451 | 33.2 | 0.37 | 15.3 | 0.1 |

Taken from Table VIII, p. 100, *Equilibrium Diagrams of Aluminium Alloy Systems* [3].

### 4.7.2   Limits of solid solubility

A knowledge of the limits of solid solubility in this alloy system is of very great importance because the wide range of solubility along the quasi-binary line, and marked decrease in solid solubility with temperature renders these alloys susceptible to heat treatment for the improvement of mechanical properties. It is therefore essential to know how the relationships are affected by an excess of magnesium or silicon over that required to form $Mg_2Si$.

In the Fig. 43 [12] isothermals have been inserted at intervals of 10°C. On the silicon side of the quasi-binary line, the apex of the three-phase triangle $(Al + Mg_2Si + Si)$ moves towards the aluminium corner with falling temperature, so that the range of compositions over which both constituents are deposited from solid solution widens as the temperature is reduced, and at 300°C extends almost to the quasi-binary line; the isothermals move closer together and approach the aluminium–magnesium face of the model. The solubility of $Mg_2Si$ falls off very rapidly in the presence of magnesium over that required to form $Mg_2Si$.

The reaction between magnesium and silicon is a balanced one, requiring an excess of either magnesium or silicon for its completion. It is possible, therefore, for free silicon to appear as a micro-constituent in alloys, the composition of which lie along the quasi-binary line, or which may contain a small excess of magnesium. The maximum divergence from the quasi-binary equilibrium occurs at about 1–2 per cent silicon where an excess of some 2 per cent magnesium is needed to suppress completely the silicon [12].

### 4.7.3  $Mg_2Si$ phase

Under the conditions of equilibrium, in alloys containing an excess of silicon, $Mg_2Si$ does not appear as a constituent of the eutectic until a magnesium content of about 0.8 per cent has been reached, but the aluminium-rich phase is incapable of reaching saturation during commercial conditions of casting, with the result that $Mg_2Si$ appears at as little as 0.2 per cent magnesium. The $Mg_2Si$ crystallizes in the 'anti-fluorite' system (see Fig. 44); the magnesium particles occupy the fluorine sites, giving a simple cubic distribution while the cube centres are alternately empty (holes) and occupied by a silicon particle or atom.

The 'holes' in $Mg_2Si$ differ from the vacancies that can exist in the lattice and from those present in $\alpha$ aluminium after heating and quenching. It can therefore be expected that there will be formed zones and precipitates or Guinier-Preston compounds near any dislocations present or formed by clustering of vacancies. Thermodynamic analysis of binary aluminium–magnesium and aluminium–silicon systems shows repulsion of aluminium–magnesium and attraction of aluminium–silicon. The precipitates or zones can thus be coherent with the matrix and produce hardening in the latter by partial coherency stressing and by dispersion.

Quenching aluminium–magnesium–silicon commercial alloys from high temperatures produces a supersaturated solution of magnesium and silicon in $\alpha$ aluminium but clusters survive since the attraction between magnesium and silicon is so high. Some of the vacancies and dislocation loops are simultaneously frozen in. On tempering at around 175°C the vacancies and

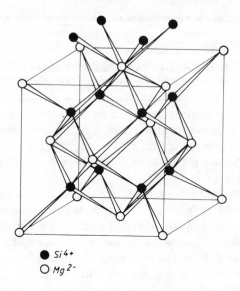

$\bullet\ Si^{4+}$

$\bigcirc\ Mg^{2-}$

Fig. 44 — 'Anti-fluoritic' structure of Mg$_2$Si.

solute atoms migrate towards the disclocation loops (or spirals) to form clusters, zones and Mg$_2$Si precipitates. This is confirmed by Hill's lattice measurements [13]. The hardening obtained after quenching and tempering is due to the dispersion of Mg$_2$Si in $\alpha$, while $\alpha$ itself is hardened by the excess of silicon or magnesium compared with the stoichiometric Mg$_2$Si ratio.

### 4.7.4 The addition of fourth metals
By adding fourth elements it is possible to improve mechanical properties of the AlMg$_2$Si alloys.

#### 4.7.4.1
In view of its coherency stress hardening effect, additions of up to 1.5 per cent copper are made to alloys of 6066 type for applications where corrosion resistance is not a prime consideration.

#### 4.7.4.2
Manganese added to ternary AlMg$_2$Si alloys produces new constituents, namely Mn$_2$SiAl$_{10}$ and (Fe,Mn)Al$_6$, though the properties are improved only slightly. However, for alloys of the 6151 type, chromium increases the yield and corrosion resistance and additions of 0.4 to 0.7 per cent manganese to alloy 6351 reduce notch sensitivity. Preferential precipitation on grain boundaries is strongly suppressed by manganese, but produces rapid preci-

pitation if the quenching rate after solution heat treatment is low, thereby lowering the final mechanical properties.

### 4.7.4.3
Chromium behaves similarly to manganese but weight for weight is twice as effective — additions of chromium are normally limited to about 0.2 or 0.3 per cent maximum.

### 4.7.5  Some uses of aluminium–magnesium–silicon alloys
Aluminium–magnesium–silicon alloys of the 6101 type are used for busbars, electrical conductors and fittings, as this type has the best combination of electrical and mechanical conductor properties, with a conductivity of 55 per cent IACS. Alloys of the 6063 type are suitable for intricate extruded sections of medium strength for architectural members such as glazing bars and window frames. The 6082-type alloy is for structural purposes, having both good strength and general corrosion resistance.

## 4.8  7XXX GROUP: ALUMINIUM–ZINC ALLOYS

### 4.8.1  Binary aluminium–zinc alloys
Aluminium and zinc form a eutectic at 95 per cent zinc, freezing at 382°C, the two constituents being a solid solution containing 82.8 per cent zinc at the eutectic temperature and zinc solid solution containing 1.14 per cent aluminium. Thus, all the aluminium-rich alloys of this system solidify as solid solutions, no other phase being capable of existence at this stage, but this solid solution breaks down at lower temperatures. From 0 to 31.6 per cent zinc it deposits zinc, the solid solubility limit rising from about 4 per cent zinc at 100°C to 31.6 per cent zinc at 275°C (see Fig. 45).

Zinc adversely affects hot shortness on solidification. Alloys of 8 to 10 per cent zinc with copper additions to improve castability are useful as general-purpose casting alloys, but corrosion resistance is not high.

Binary aluminium–zinc alloys do not find general application as wrought products but some sheet of 1 per cent zinc 7072 is used for cladding a range of alloys, such as 3003, 3004, 5050, 5150, 6061, 7075 and 7178. With these exceptions the binary alloys are not of commercial interest but the aluminium–zinc–magnesium derivatives are high strength alloys used extensively in aircraft construction.

### 4.8.2  Aluminium–zinc–copper
In this system of alloys the primary aluminium base field is very large, extending from 33 per cent copper at 0 per cent zinc and 5 per cent copper at 90 per cent zinc. The primary aluminium field is bounded by four valleys, all

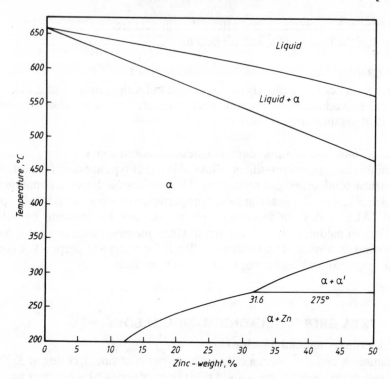

Fig. 45 — Aluminium–zinc.

eutectic in character, associated with the simultaneous separation of aluminium with $CuAl_2$, T phase (an intermetallic constituent of the copper–zinc system) and zinc. The invariant points at which these valleys meet are listed in Table 20.

For alloys of aluminium having low copper and moderate zinc content the liquidus surface is sensibly plane (Fig. 46). The solidus surface over the same limited range of compositions is given in Fig. 47. Only one of the two phase fields occurs, that in which solidification is completed during the simultaneous separation of aluminium and $CuAl_2$.

With falling temperature, the aluminium-rich apices of the various three-phase triangles move from the invariant points of the solidus surface towards the aluminium corner, but only one of these, the apex of the triangle Al–$CuAl_2$–T, enters the range of compositions covered by the diagram Fig. 48, and that only when the temperature has fallen to 320°C. Isothermal points have been inserted at 10°C intervals along the locus of the apex, but the three-phase triangles at intervals of 20°C only.

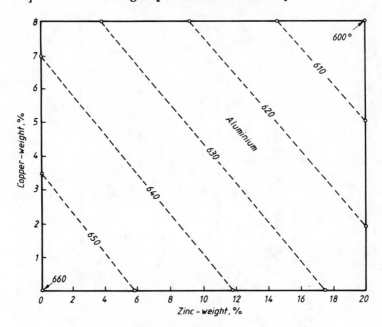

Fig. 46 — Aluminium–copper–zinc liquidus surface.

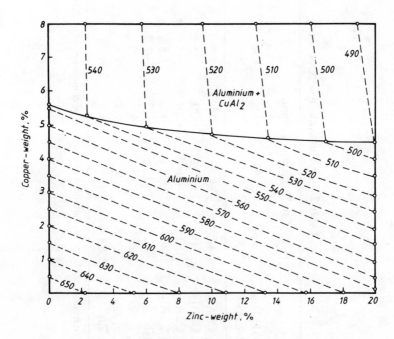

Fig. 47 — Aluminium–copper–zinc solidus surface.

**Table 20** — Aluminium–copper–zinc — boundaries of primary aluminium field

| Solid phases present | Temp °C | Composition of liquid | | Composition of solid phase | | | | | | | | |
|---|---|---|---|---|---|---|---|---|---|---|---|
| | | | | Al | | T | | ε | | Zn | |
| | | Cu% | Zn% | Cu % | Zn % | Cu % | Zn % | Cu % | Zn % | Al % | Cu % |
| Al, CuAl$_2$ | 548 | 33 | 0 | 5.7 | 0 | — | — | — | — | — | — |
| Al, CuAl$_2$ | 500 | ~29 | ~20 | ~5 | ~17 | — | — | — | — | — | — |
| Al, CuAl$_2$ | 450 | ~20 | ~48 | ~3 | ~32 | — | — | — | — | — | — |
| Al, CuAl$_2$→T | 420 | 15 | 60 | 1.5 | 65 | 55.5 | 13 | — | — | — | — |
| Al, T | 400 | ~11.5 | ~70 | ~1.7 | ~68 | 55.5 | 14 | — | — | — | — |
| Al, T→ε | 396 | 10.5 | 74 | 1.8 | 72 | — | — | 23 | 72 | — | — |
| Al, ε, Zn | 379.5 | 3.71 | 89.37 | 1.5 | 78.1 | — | — | 15.5 | 83.3 | 1.25 | 2.75 |
| Al, Zn | 382 | 0 | 95 | 0 | 82.8 | — | — | — | — | 1.14 | 0 |

Taken from Table V, p. 74, *Equilibrium Diagrams of Aluminium Alloy Systems* [3].

### 4.8.3   Aluminium–zinc–magnesium

In the ternary aluminium–zinc–magnesium equilibrium diagram the primary aluminium phase field extends as far as 35 per cent Mg, where it terminates in a eutectic at 450°C with $Mg_2Al_3$. It extends along the zinc axis to 95 per cent zinc, the binary eutectic with zinc at 382°C. From these two eutectics, binary valleys extend into the diagram, the former terminating in an invariant point at 30 per cent magnesium, 13 per cent zinc and 447°C (where the third solid phase is a ternary phase designated T) and the latter in a ternary eutectic point of aluminium, zinc and $MgZn_5$ at 342–8°C and a composition 2.95 per cent magnesium, 93 per cent zinc. The constituent T exists over a wide homoeneity range extending from 33 per cent magnesium, 23 per cent zinc to 22 per cent magnesium, 64 per cent zinc. It forms a quasi-binary system with aluminium which comprises a series of alloys in which the magnesium content is 0.37 times that of zinc. There is a eutectic at 18 per cent magnesium, 45.6 per cent zinc freezing at 489°C, which lies at the highest point of the eutectic valley separating the aluminium and T primary fields. On the magnesium side of the quasi-binary line the alloys comprise a partial system in which the only constituents that occur are Al, $Mg_2Al_3$ and T. On the zinc side, the Al–T valley runs down to a eutectic at 475°C, the constituents being Al, T and $MgZn_2$, with the composition 11.3 per cent magnesium, 60.4 per cent zinc. Very close to this point a second quasi-binary line is crossed, Al–$MgZn_2$, corresponding with a series of alloys in which the magnesium content is 0.2 times that of the zinc. This sets up a second partial ternary system, with constituents Al, T and $MgZn_2$. The alloys, still richer in zinc, form a third partial system in which the Al–$MgZn_2$ binary valley runs down to a peritectic point (365°C, 3 per cent Mg, 91 per cent Zn) where $MgZn_5$ is involved, and from this point the Al–$MgZn_5$ valley runs down to the ternary eutectic.

*Liquidus surface*
The liquidus surface of the aluminium-rich alloys of commercial interest consists of a single surface, due to aluminium, since the binary valleys lie well outside the range of these compositions.

*Solidus surface*
The solidus surface of the aluminium-rich alloys with zinc and magnesium is shown in Fig. 49. The field, bearing curved isothermals, which extends along the aluminium–zinc axis and as far as 15.35 per cent along the aluminium–magnesium axis, is due to aluminium alone; its upper boundary contains three invariant points viz.

(a) The apex of the three-phase triangle at 447°C of which the phases are

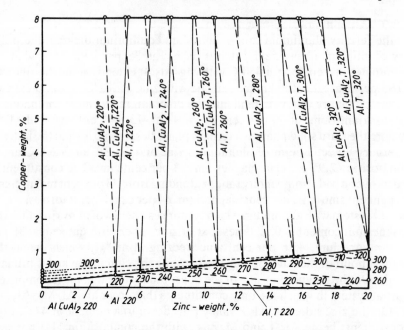

Fig. 48 — Aluminium–copper–zinc: limits of solid solubility.

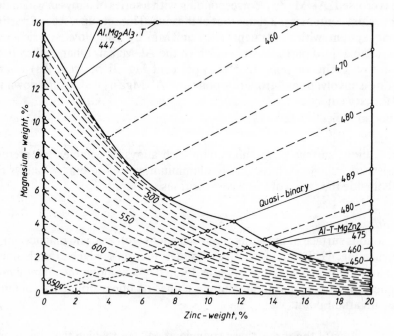

Fig. 49 — Aluminium–magnesium–zinc solidus.

aluminium containing 12.3 per cent magnesium, 1.8 per cent zinc; $Mg_2Al_3$ containing 33.5 per cent magnesium, 10 per cent zinc; and T containing 30 per cent magnesium, 26 per cent zinc.

(b) The point of intersection with the quasi-binary section Al–T, the aluminium here containing 4.3 per cent magnesium, 11.6 per cent zinc; and T having 21 per cent magnesium, 5.4 per cent zinc at the temperature 489°C.

(c) The apex of the three-phase triangle Al–T–$MgZn_2$ at 475°C, the aluminium containing 2.8 per cent magnesium, 14.3 per cent zinc; T having 20 per cent magnesium, 64 per cent zinc; and $MgZn_2$ with 15.5 per cent magnesium, 82.6 per cent zinc (the last with a little dissolved aluminium of the order of 2 per cent).

Of the two phase fields of the solidus, that due to aluminium and $Mg_2Al_3$ is almost horizontal; that due to aluminium and T is ridge-shaped, having its crest along the quasi-binary line, while that due to aluminium and $MgZn_2$ slopes fairly rapidly.

### 4.8.3.1   Limits of solid solubility

With falling temperature, the apices of the two three-phase triangles Al–$Mg_2Al_3$–T and Al–T–$MgZn_2$ move towards the aluminium corner. Their loci have been plotted as dotted lines in Fig. 50 with ringed points at appropriate temperature intervals. The sides of the three-phase triangles, meeting at the ringed points, have been drawn as broken lines. The solid solubilities along the quasi-binary line Al–T have also been ringed.

### 4.8.3.2   Failure to reach Equilibrium

It is certain that the aluminium-rich solid solution will not reach saturation during solidification, so that along the quasi-binary line Al–T, the latter constituent will appear in the microsections at low percentages of zinc and magnesium. Again, in the partial system Al–T–$Mg_2Al_3$, failure to reach saturation is bound to lead to the appearance of $Mg_2Al_3$ over a wide range of compositions, possibly extending down to a line running from 3.5 per cent magnesium or 0 per cent zinc to the composition of T and therefore parallel to the 200°C solid solubility isotherm. In the second partial system Al–T–$MgZn_2$, failure to reach saturation in the aluminium-rich phase will necessarily result in the premature appearance of T and $MgZn_2$ but no precise figure can be given.

### 4.8.4   The effect of iron

Iron, if present alone as an impurity, would occur solely as $FeAl_3$. In ternary alloys of aluminium–magnesium–iron, the primary $FeAl_3$ field is entered at lower percentages of iron as the magnesium content is increased; and

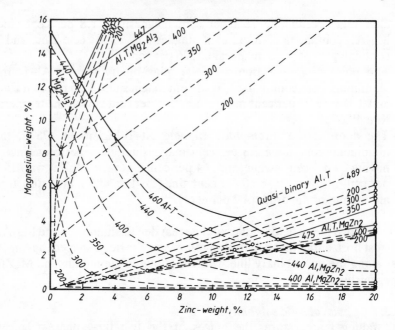

Fig. 50 — Aluminium–magnesium–zinc: limits of solid solubility.

similarly in ternary alloys of aluminium–iron–zinc, as also in the simultaneous presence of magnesium and zinc.

### 4.8.5    The effect of silicon

Silicon, if present alone, combines with magnesium to form $Mg_2Si$ and this occurs whatever the ratio of magnesium to zinc. The simultaneous presence of iron and silicon results in the occurrence of two additional constituents α (FeSi) and β (FeSi), both formed peritectically from $FeAl_3$. The retention of $FeAl_3$ is favoured by a high magnesium/zinc ratio, and a rapid rate of solidification which covers the suppression, in whole or in part, of the two peritectic reactions. Silicon contents in excess of 0.3 per cent reduce mechanical properties.

### 4.8.6    Hardening of aluminium–zinc–magnesium alloys

Alloys of the aluminium–zinc–magnesium series harden by the dispersion effect of transition intermetallics. On quenching and ageing at 120 to 160°C, a series of stages is passed through to result in the formation of the Guinier-Preston compound (designated 'M' for '$MgZn_2$'), the lattice of which is well defined.

Preferential precipitation at the boundaries between grains and subgrains results in a drop in magnesium and zinc concentration in the vicinity of boundaries. Regions free from zones or platelets of precipitate can be found around the grain boundaries (thickness about 0.5 Å) and are responsible for the technological faults of brittleness and intergranular corrosion encountered in Al–Zn–Mg alloys. The development of more complex compositions and thermal treatments coupled with changes in processing have led to advances in combinations of resistance to stress corrosion and exfoliation corrosion, high mechanical properties, higher fracture toughness and good fatigue strength. Elements such as copper, chromium and/or manganese [14] are added to aluminium–zinc–magnesium alloys to control the mode of precipitated phases and the nature of the grain boundary structure. Additions of silver have also been found to be beneficial [15] in this context but have been ruled out by considerations of cost and of availability in adequate quantities.

### 4.8.7   Effect of other elements
#### 4.8.7.1
Additions of 0.5 per cent copper to the ternary aluminium–zinc–magnesium alloys are sufficient to influence the transformation kinetics; the rate of hardening by zone formation is reduced compared with that of ternary alloys. This is attributed to the formation of more complex Al–Zn–Mg–Cu zones than in the Al–Zn–Mg system. However, the presence of manganese appears to eliminate the retarding influence of copper, and this is possibly due to manganese taking some of the copper into solid solution.

#### 4.8.7.2
Thomas and Nutting [16] report that manganese appears to accelerate the precipitation of $Mg'Zn_2(M')$ from Guinier-Preston zones. Manganese also prevents coarse-grained recrystallization structures in quenched extruded products.

#### 4.8.7.3
Chromium may be added to improve stress corrosion resistance either singly or in conjunction with manganese. There is no known compound between chromium and magnesium but it can take about 30 per cent zinc into solid solution and also forms $CrZn_{17}$. With aluminium there is a monoclinic compound $CrAl_7$ in which manganese can dissolve. The solubility of chromium in aluminium is only 0.015 per cent at 300°C, much lower than that of manganese, and any added chromium will consequently readily precipitate in the dislocations and grain boundaries before zones are formed there. Chromium additions in the range 0.1–0.4 per cent promote a highly uniform zone distribution inside the matrix. Copper promotes the formation of $CrAl_7$ in the grain boundaries.

**4.8.8   Cast alloys of aluminium–zinc–copper–magnesium**
Cast alloys of the quaternary compositions with zinc–copper and magnesium are more hot short than those of the copper or silicon series; they require high purity aluminium as a base in order to develop high mechanical properties without the need for heat treatment; they braze readily, giving good mechanical properties, and anodise to an excellent finish.

Aluminium–zinc–magnesium–copper alloys in the forms of plate, sheet, extrusions and forgings have been used in large quantities over a number of years [17].

The most commonly used alloy of the series has been 7075 and variants of the basic composition of this alloy. 7175 and 7475 with lower impurity limits and adjusted manganese and chromium are used for parts of large cross-section to meet more stringent requirements for elongation in directions perpendicular to that of major extension in working. Increasing base metal purity also improves fracture toughness.

Alloy 7049 has higher zinc and lower chromium than 7075. The lower chromium is to reduce quench sensitivity. Alloy 7149 has 0.2 per cent iron and 0.15 per cent maximum silicon to improve ductility and fracture toughness.

Alloy 7050 was designed with particular reference to thicker section parts. This alloy has higher zinc and copper contents than 7075 and has zirconium in place of chromium, both to suppress recrystallization and to reduce the strength loss with decreasing quench cooling rates. The $ZrAl_3$ phase avoids combination with principal alloying species; it is present in small volume fraction as compact nearly-spherical particles, factors favouring high fracture toughness.

## 4.9   8XXX GROUP: OTHER ELEMENT ALLOYS
This is a miscellaneous group of alloys.

**4.9.1   Aluminium–iron–silicon less than 99.00 per cent aluminium**
Wrought aluminium–iron–silicon alloys having an aluminium content of less than 99.00 per cent fall in the 8XXX series. These alloys have higher mechanical strength than those of higher purity.

**4.9.2   Aluminium–nickel–iron alloy**
Alloy 8001 is a low-nickel, low-iron–aluminium alloy and is of interest from the point of view of aqueous corrosion at temperatures above 100°C.

### 4.9.3 Aluminium–tin alloys

Alloy 8020 has tin as the major alloying element together with small additions of silicon, iron, copper and nickel. Alloys of this type are used as aluminium-alloy-bearing metals.

### 4.9.4 Aluminium–lithium alloys

These alloys, offering predicted weight savings of 10–15 per cent, are likely to have considerable impact on the future design, manufacture and economies of aircraft [18].

The binary alloy system is given in Fig. 51, and a series of alloys of special interest for aircraft purposes have been developed. They are based upon

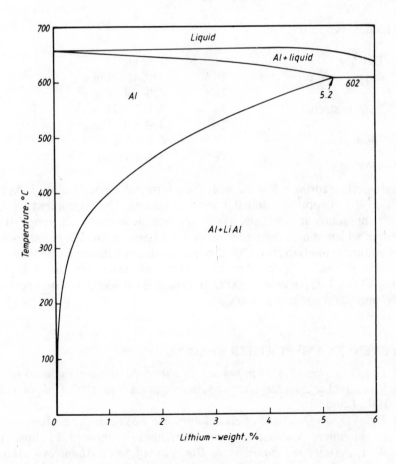

Fig. 51 — Aluminium–lithium.

additions of 2–3 per cent lithium with copper and magnesium. Their density is about 10 per cent lower than that of other alloys and the value of $E$ is 10 per cent higher, making them particularly useful as structural materials. In the UK they are the subject of DTD Specifications (aircraft alloys). Composition (DTD XXXX) is.

| | |
|---|---|
| Lithium | 2.30–2.60 |
| Copper | 1–1.40 |
| Magnesium | 0.5 –0.90 |
| Silicon | 0.20 max |
| Iron | 0.30 max |
| Zirconium | 0.10–0.14 |
| Sodium | 0.002 max |

Mechanical Properties are

| | | |
|---|---|---|
| Density | | 2.52 g/cm$^3$ |
| 0.2 per cent proof stress | T4 | 370–420 N/mm$^2$ |
| | T6 | 420–480 N/mm$^2$ |
| Tensile strength | T4 | 440–500 N/mm$^2$ |
| | T6 | 490–540 N/mm$^2$ |
| Elongation | T4 | 6 per cent |
| | T6 | 5–6 per cent |

Aluminium forms a eutectic with the intermetallic constituent Li–Al at 602°C [19] and approximately 9.9 per cent lithium. The constituent Li–Al melts congruently at 718°C and has an appreciable homogeneity range, the extent of which at high temperatures is not accurately known, but at room temperature extends to about 20–22 per cent lithium. Lithium is appreciably soluble in aluminium up to 5.2 per cent at the eutectic temperature, 2.55 per cent at 500°C, 1.02 per cent at 400°C, 0.32 per cent at 300°C, 0.06 per cent at 200°C and 0.005 per cent at 100°C.

## REFERENCES AND FURTHER READING

[1] *The Properties of Aluminium and its alloys,* Aluminium Federation.
[2] Van Lancker, M., *Metallurgy of Aluminium Alloys,* 1967, Chapman & Hall, London.
[3] *Equilibrium Diagrams of Aluminium Alloy Systems,* The Aluminium Development Association (now Aluminium Federation), Phillips, H. W. L., *Annotated Equilibrium Diagrams of Some Aluminium Alloy Systems,* Institute of Metals, Monograph No. 25.
[4] For example,Guinier-Preston zones, *JIM,* Aug. 1959, **87,** 430.

[5] Phragmen, G., *JIM,* 1950, **77,** 489.

[6] Kralik, *Hutnicky Listy,* 1958, **13,** pp. 49–51.

[7] Raynor, G. V., *Annotated Equilibrium Diagrams,* Institute of Metals, Monograph No. 5, 1945.

[8] Little, G. T., Raynor, G. V. and Hume-Rothery, W., *J. Inst. Metals,* 1943, **69,** 423.
Butcher, E., Raynor, G. V. and Hume-Rothery, W., *ibid.,* 1943, **69,** 209.
Butcher, E. and Hume-Rothery, W., *ibid.,* 1948, **71,** 291.
Wakeman, D. W. and Raynor, G. V., *ibid.,* 1948–49, **75,** 131.

[9] Ransley, C. E. and Talbot, D. E. J., *J. Inst. Metals,* 1959–60, **88,** 150–158.

[10] Ransley, C. E., British Patent 823, p. 802.

[11] Ransley, C. E., Talbot, D. E. J., Beton, R. H. and Bryant, A. J., *JIM Discussion,* 1960–61, **89,** 316–318.

[12] Phillips, H. W. L., *JIM,* 1946, **72,** 151.

[13] Hill, R. B. and Axon, H. J. *JIM,* 1955, **83,** 354.

[14] Rosenkranz, W., *Aluminium,* 1960, **36,** 250–257 and 395–408.

[15] Polmear, I. J., *JIM,* 1960–61, **89,** 51–59 and 193–202.

[16] Thomas and Nutting, *JIM,* 1959, **88,** 81.

[17] Hunsicker, H. Y., *Phil. Trans. R. Soc., London,* **282,** 359–376.

[18] Grimes, R., Cornish, A. J., Miller, W. S. and Reynolds, M. A., *Metals and Materials,* June 1985, **2,** p. 357.

[19] Nowak, S. K., *T. Amer. Inst. Min. and Met. Eng.,* 1956, **206**(5), 553.

[20] Private communication, D. Lewis.

# 5

# Properties of aluminium alloys

## 5.1 GENERAL

Aluminium and its alloys are divided into three broad classes, ingot for remelting, shaped castings and mechanically worked (wrought) products. The last-mentioned class is subdivided into work hardening (non-heat treatable) and heat treatable alloys and into the various forms produced by mechanical working, in particular re-roll stock, plate, sheet, foil, bar, extrusions, hollow sections, forging stock, forgings, tube, wire, rivet and bolt stock, and solid conductor.

Aluminium powder is used in a number of applications and broadly it can be fitted into two classes: powder particles produced by atomising molten metal which are roughly equiaxed; and flake powder which is produced mainly by treating atomised powder, or foil, in a ball mill to yield flakes with a large diameter-to-thickness ratio.

### 5.1.1 Ingot

Ingot is a term applied to solid aluminium or aluminium alloy in the unwrought form in a wide variety of shapes and sizes depending upon the intended use. Ingot for remelting is designed to suit the handling and melting equipment, that for plate or sheet is rectangular and that for extrusion is normally cylindrical.

The impurity content of ingot for remelting made direct from the reduction plant will generally be much lower than that of reclaimed metal. The chemical composition limits for ingot for remelting are governed by the product for which it is to be used.

The physical property of an ingot for remelting which is of most interest is the melting temperature. A low melting temperature is desirable since it enables the final alloy to be melted without the need for using excessively high temperatures, hence the use of master alloys. Manganese has a melting temperature of 1244°C whereas an aluminium–10 per cent manganese master alloy melts at a temperature of about 790°C. Copper melts at 1083°C and an aluminium–33 per cent copper master alloy melts at 548°C (see Fig. 29). The subject of melting and alloying is dealt with more fully in Chapter 6.

### 5.1.2 Castings [1]

The three most commonly used processes are sand casting, permanent mould casting and die casting. Sand moulds are fed with molten metal by gravity. The metal moulds used in permanent mould casting are fed either by gravity or by using low pressure air or other gas to force metal up a sprue (channel) into the mould; electromagnetic pumps may also be used for this purpose. When sand cores are used in permanent moulds the process is sometimes called semi-permanent mould casting. In die casting, molten metal is forced into the steel die at high presure by the action of one or more hydraulic rams.

Alloys used for sand and permanent mould castings are frequently chosen from compositions which respond to solution heat treatment and quenching to yield components with relatively high strength and good machining characteristics. Because of their gas content, die castings are not generally suitable for heat treatment unless special casting techniques, such as vacuum die casting, the Japanese patented 'pore free' die casting system or the 'Accurad' system developed by General Motors are employed. The process produces components with comparatively high mechanical properties and very close dimensional tolerances [1].

### 5.1.3 Wrought aluminium alloy products

The wrought alloys of aluminium may be divided into two main groups.

#### 5.1.3.1 Work hardening alloys

These are alloys which depend upon deformation by rolling, wire drawing, tube drawing, etc., for the development of structures and properties required for commercial applications.

#### 5.1.3.2 Heat treatable alloys

After working to develop the required shape as plate, extrusions, forgings, wire, tube, etc., these alloys are then subjected to thermal treatment at elevated temperatures to take alloying constituents into solid solution. The metal is quenched from the solution heat treatment temperature. Depending upon the composition and intended use, the alloy in the quenched condition is subjected to straightening and flattening operations and the desired level of mechanical properties is developed by allowing the alloy to naturally age at ambient temperature for a period of a few days or is subjected to a precipitation treatment (artificial ageing) at a temperature usually within the range 120–200°C. (See Chapter 6.)

## 5.2 CASTING ALLOYS

### 5.2.1 Casting alloy chemical compositions
The alloying element content of a number of general engineering and aerospace casting alloys is given in Table 21.

### 5.2.2 Casting alloy designations
In the UK, casting alloys for general engineering are specified in BS 1490 and are numbered LM0 to LM31, the letters LM originally signifying Light Metal. Over the years, several alloys in this series have been withdrawn. In Table 21 the alloys have been arranged in an order based on the amount of the major alloying contents of silicon, copper and magnesium. This order is maintained in Tables 22–26 to facilitate comparisons between alloys of different compositions.

For castings for aerospace purposes, there are several British Standards in the L series and some DTD Specifications with casting alloy heat treatment designations. The various suffixes listed below indicate the condition of the casting;

    M  = As cast
    TB  = Solution heat treated and naturally aged
    TB7 = Solution heat treated and stabilised
    TE  = Artificially aged (precipitation treated)
    TF  = Solution heat treated and artificially aged
    TF7 = Solution heat treated, artificially aged and stabilised.
    TS  = Thermally stress relieved

### 5.2.3 Typical properties and characteristics of casting alloys
#### 5.2.3.1 Casting characteristics
The casting characteristics of British Standard General Engineering Aluminium Alloys are given in Table 23. It will be seen that unalloyed aluminium is only rated as having 'Fair' casting properties; this is accounted for by the fact that its resistance to hot tearing is poor, that is, the stresses set up as the metal solidifies and shrinks are sufficient to cause fractures at a high temperature.

As the silicon content of the alloys increases from 3 to 5 per cent up to 10–13 per cent, the fluidity and resistance to hot tearing improves from

Table 21 — Castings — basic chemical compositions of General Engineering BS 'L' and DTD Specifications

| Material designation | | Chemical composition wt % | | |
| --- | --- | --- | --- | --- |
| General engineering BS | Aerospace BS L/DTD | Si | Cu | Other essential elements† |
| LM0 | | | | 99.50 min Al |
| LM18 | | 4.5–6.0 | | Mg |
| | DTD 716B | 3.5–6.0 | | |
| | DTD 722B | | | |
| | DTD 727B | | | |
| | DTD 735B | | | |
| LM16 | 3L78 | 4.5–5.5 | 1.0–1.5 | Mg |
| LM22 | | 4.0–6.0 | 2.8–3.8 | Mn |
| LM4 | | 4.0–6.0 | 2.0–4.0 | Mn |
| LM25 | 2L99 | 6.5–7.5 | | Mg |
| LM27 | | 6.0–8.0 | 1.5–2.5 | Mn |
| LM21 | | 5.0–7.0 | 3.0–5.0 | Mg, Mn |
| LM26 | | 8.5–10.5 | 2.0–4.0 | Mg |
| LM6 | | 10.0–13.0 | | Mg, Mn |
| LM9 | | 10.0–13.0 | | |
| LM20 | | 10.0–13.0 | | |
| LM2 | | 9.0–11.5 | 0.7–2.5 | |
| LM13 | | 10.0–12.0 | 0.7–1.5 | Mg |
| LM30 | | 16–18 | 4.0–5.0 | Mg |
| LM28 | | 17–20 | 1.3–1.8 | Mg, Ni |
| LM29 | | 22–25 | 0.8–1.3 | Mg, Ni |
| | 2L91 | | 4.0–5.0 | |
| | 2L92 | | | |
| | L154 | | | |
| | L155 | 1.0–1.5 | 3.8–4.5 | Ti |
| | L119 | | 4.5–5.5 | Ni, Ti, Cr, Co, Sb |
| | 4L35 | | 3.5–4.5 | Mg, Ni |
| LM12 | | | 9.0–11.0 | Mg |
| LM5 | | | | Mg, Mn |
| | 5018A | | | Zn |
| LM10 | 4L53 | | | Mg, Mn |
| LM31 | DTD 5008B | | | Zn, Ti, Cr |
| | 3L51 | 1.5–2.8 | 0.8–2.0 | Mg, Ni, Fe |
| | 3L52 | 0.6–2.0 | | Mg, Ni, Fe |

† Reference must be made to the complete British Standard or DTD Specifications for full details of the essential elements and impurity levels.

'Good' to 'Excellent' and so does the suitability for casting, with the exception of LM24 which is only 'Fair' for sand casting and permanent mould casting but this alloy, LM24, is 'Excellent' for die casting whereas other alloys except LM4, LM6, LM20 and LM30 are not usually used for this method of casting. The high silicon alloys are rated 'Poor' for sand casting and are only 'Fair' for permanent mould casting.

**Table 22** — Casting characteristics of BS 1490 General Engineering
Aluminium Alloys

| Alloy designation | Suitability for | | | Casting characteristics | |
|---|---|---|---|---|---|
| | Sand casting | Permanent mould casting | Die casting | Fluidity | Resistance to hot tearing |
| LM0 | F | F | F† | F | P |
| LM18 | G | G | G† | G | E |
| LM16 | G | G | G† | G | G |
| LM22 | G† | G | G† | G | G |
| LM4 | G | G | G | G | G |
| LM25 | G | E | G† | G | G |
| LM27 | G | E | G† | G | G |
| LM21 | G | G | G† | G | G |
| LM26 | G | G | F† | G | G |
| LM24 | F† | F† | E | G | G |
| LM6 | E | E | G | E | E |
| LM9 | G | E | G† | G | E |
| LM20 | E† | E | G | E | E |
| LM2 | G† | G† | E | G | E |
| LM30 | — | F | G | G | G |
| LM28 | P | F | — | F | G |
| LM29 | P | F | — | F | G |
| LM12 | F | G | U | F | G |
| LM5 | F | F | F† | F | F |
| LM10 LM31 | F | F | F† | F | G |

† Not usually cast by this method.
E=Excellent. G=Good. F=Fair. P=Poor. U=Unsuitable.

The 10 per cent copper alloy is 'Good' for permanent mould casting. The 5 per cent and 10 per cent magnesium are rated as having 'Fair' casting characteristics.

The alloys LM6, LM18 and LM20 give 'Excellent' pressure tightness and the other alloys having 3.5–13 per cent silicon content are generally rated as 'Good' whilst the high 17–25 per cent silicon alloys are considered to give 'Fair' only pressure tightness.

**Table 23** — Casting alloys — general properties

| Alloy designation | Resistance for atmospheric attack | Machinability | Pressure tightness | Plating | Vitreous enamelling | Protective | Anodising Colour | Bright |
|---|---|---|---|---|---|---|---|---|
| | | | | | | Suitability for | | |
| LM0 | E | F | F | E | E | E | E | E |
| LM18 | E | F | E | F | E | G | F(D) | U |
| LM16 | G | G | G | F | — | G | F(D) | U |
| LM22 | G | G | G | G | — | G | F | U |
| LM4 | G | G | G | G | — | G | F(D) | U |
| LM25 | E | F | G | F | — | G | F(D) | U |
| LM27 | G | G | G | G | — | G | F(D) | U |
| LM21 | G | G | G | F | — | F | U | U |
| LM26 | G | F | F | F | — | F | U | U |
| LM24 | G | F | G | F | — | F | F(D) | U |
| LM6 | E | F | E | F | E | F | U | U |
| LM9 | E | F | G | F | — | F | U | U |
| LM20 | G | F | E | G | E | G | F(D) | U |
| LM2 | G | F | G | F | — | F | U | U |
| LM30 | G | P | F | F | — | U | U | U |
| LM28 | G | P | F | F | — | U | U | U |
| LM29 | G | P | F | F | — | U | U | U |
| LM12 | P | E | G | G | — | F | F | U |
| LM5 | E | G | P | F | U | E | E | G |
| LM10 | E | G | P | U | U | E | F | U |

E=Excellent. G=Good. F=Fair. P=Poor. U=Unsuitable. D=Dark colours only.

### 5.2.3.2 Casting alloys: general properties

With the exception of the 16–25 per cent silicon alloys which have poor machinability, attributable to the presence of large, hard particles of primary silicon, most of the other alloys give 'Fair' to 'Good' machinability improving with increase in copper content, the 9–11 per cent copper alloy LM12 giving 'Excellent' machinability.

The resistance to atmospheric attack of the listed LM casting alloys in Table 23 is 'Good' to 'Excellent'.

The suitability of unalloyed aluminium castings for electroplating is 'Excellent', whereas that of casting alloys in general is considered to be 'Fair' and the 10 per cent magnesium alloy, LM10, is unsuitable for plating. The 10 per cent copper alloy, LM12, has good plating characteristics, but all castings to be plated must be free from casting defects, on the surface, such as pores, cracks or blisters.

Unalloyed aluminium, LM0, and the binary aluminium 4.5–6.0 per cent and 10.0–13.0 per cent silicon alloys, LM18 and LM6, have 'Excellent' suitability for vitreous enamelling as also has LM20 which differs from LM18 in that it has a higher iron content of up to 1 per cent. The high magnesium alloys, LM5 and LM10, are unsuitable for vitreous enamelling. No data are available on the other alloys in Table 23.

The suitability of the silicon-bearing alloys for protective anodising decreases with increasing silicon content, the 17–25 per cent silicon alloys being unsuitable for this process. The 10 per cent copper alloy gives 'Fair' results and the unalloyed LM0 and the 5 and 10 per cent magnesium alloys are 'Excellent' when anodised for protection. The anodised film on LM0 and LM5 is 'Excellent' for colouring. The silicon-bearing alloys give a grey to black anodic film and are suitable for dark colours only, those with the higher silicon contents being 'Unsuitable' for colouring.

Most of the casting alloys are unsuitable for bright anodising, the exceptions being LM0 and LM5 which are 'Excellent' and 'Good' respectively.

### 5.2.4 Physical properties of casting alloys
The physical properties of a number of casting alloys are given in Table 24

**Table 24** — Physical properties of some aluminium casting alloys (BS 1490)

| Material designation BS 1490 | Temper | Density g/cm³ | Coefficient of linear expansion 200–100 °C 10⁻⁹/K | Thermal conductivity at 25°C W/mK | %IACS | Resistivity at 20°C μΩ cm | Conductivity at 20°C %IACS |
|---|---|---|---|---|---|---|---|
| LM0 | M | 2.70 | 24 | 209 | 53.1 | 3.02 | 57 |
| LM18 | M | 2.69 | 22 | 142 | 36.1 | 4.65 | 37 |
| LM25 | M | 2,68 | 22 | 151 | 38.4 | 4.4 | 39 |
| LM6 | M | 2.65 | 20 | 142 | 36.1 | 4.65 | 37 |
| LM28 | TE | 2.68 | 18 | 134 | 34.0 | — | — |
| LM29 | TF | 2.65 | 16 | 126 | 32.5 | — | — |
| LM12 | M | 2.94 | 22 | 130 | 33.0 | 5.23 | 33 |
| LM5 | M | 2.65 | 23 | 138 | 35.1 | 5.56 | 31 |
| LM10 | TB | 2.57 | 25 | 87.9 | 22.3 | 8.62 | 20 |

#### 5.2.4.1 Density
The density range of representatives of the BS 1490 series of alloys is from 2.57 g/cm³ for LM10 increasing to 2.94 g/cm³ for LM12. The density values decrease from 2.70 g/cm³ for LM0 down to 2.65 g/cm³ with increase in silicon content. The effect of magnesium in reducing density is much greater than that of silicon. From the values for LM0, LM5 and LM10, 0, 5 and 10 per cent of magnesium, the densities of 2.70, 2.65 and 2.57 respectively, it appears that the reduction is not linear with magnesium content.

### 5.2.4.2  *Expansion*

Increasing silicon contents has the effect of reducing the coefficient of linear expansion of aluminium castings from $24 \times 10^{-6}/°C$ for LM0 to $16 \times 10^{-6}/°C$ for LM9. The values for the high magnesium alloys, LM5 and LM10, are very close to those of LM0 and there is insufficient information to confirm any definite trend. The high copper alloy has a value slightly lower than that of LM0.

### 5.2.4.3  *Conductivities — thermal and electrical*

Expressed as W/m°C, the thermal conductivity of the silicon series of casting alloys decreases from 209 to 126 with increasing silicon contents from 0.3 per cent maximum up to 22–25 per cent. The thermal conductivity for 10 per cent copper is 130 W/m°C and that of the 10 per cent magnesium alloy, LM10, is much lower at 87.9 W/m°C.

As with thermal conductivity the effect of increasing silicon content is to reduce electrical conductivity values. The effect is not linear, the electrical conductivity values for 5 per cent LM18 and 12 per cent LM6 being of the order 37 per cent compared with 57 per cent IACS (International Annealed Copper Standard) for LM0. The value for 10 per cent copper alloy is down to 33 per cent IACS. The effect of increasing magnesium content to 5 per cent and 10 per cent is to reduce electrical conductivity to 31 per cent and 20 per cent IACS respectively.

### 5.2.4.4  *Elastic modulus*

A mean value of Young's modulus of elasticity for the casting alloys is $60 \times 10^3 \, \text{N/mm}^2$.

### 5.2.5  Mechanical properties of casting alloys

With castings the ability of an alloy to meet specified requirements is demonstrated by making tensile tests on test pieces machined from separately cast test bars. For general engineering the forms of these bars are specified on BS 1490. The fact that the test specimen meets specification requirements does not necessarily mean that the castings it represents will also yield the same level of properties irrespective of the section thickness and location. Information of this type can be obtained only by destructive testing of selected castings.

The specification minimum tensile test requirements for a number of casting alloys are given in Table 25 which illustrates that chill cast material has tensile properties some 50 N/mm$^2$ higher than those for sand cast. The elongation values are slightly higher for the chill castings. Elongation is generally low for material in the M condition, being improved by solution heat treatment, where appropriate, which in the case of alloys such as LM4 and LM25 improves tensile strength by some 120 N/mm$^2$.

**Table 25** — Mechanical properties of casting alloys to general engineering and aerospace specifications

| Material designation | Condition | Sand cast | | | | Chill cast | | | |
|---|---|---|---|---|---|---|---|---|---|
| | | 0.2% proof stress N/mm² | Tensile strength N/mm² | Elongation % min | Hardness Brinell | 0.2% proof stress N/mm² | Tensile strength N/mm² | Elongation % min | Hardness Brinell |
| LM0 | M | — | — | — | — | — | — | — | — |
| LM18 | M | | 120 | 3 | | | 140 | 4 | |
| DTD 735B | TF | 215† | 230 | | | 215† | 278 | 2 | |
| LM16 | TB | | 170 | 2 | | | 230 | 3 | |
| | TF | | 230 | — | | | 280 | — | |
| LM22 | TB | | — | | | | 245 | 8 | |
| LM4 | M | | 140 | 2 | | | 160 | 2 | |
| | TF | | 230 | — | | | 280 | — | |
| 2L99 | TF | 185 | 230 | 2 | | 200 | 280 | 5 | |
| LM25 | M | | 130 | 2 | | | 160 | 3 | |
| | TF | | 230 | — | | | 280 | 2 | |
| LM27 | M | | 140 | 1 | | | 160 | 2 | |
| LM21 | M | | 150 | 1 | | | 170 | 1 | |
| LM26 | TE | | — | — | | | 210 | — | 90–120 |
| LM24 | M | | — | — | | | 180 | 1.5 | |
| LM6 | M | | 160 | 5 | | | 190 | 7 | |
| LM9 | M | | — | — | | | 190 | 3 | |
| | TE | | 170 | 1.5 | | | 230 | 2 | |
| | TF | | 240 | — | | | 295 | — | |
| LM20 | M | | — | — | | | 190 | 5 | — |
| LM2 | M | | — | — | | | 150 | — | |
| LM13 | TE | | — | — | | | 210 | — | 90–120 |
| | TF7 | | 170 | — | 100–150 | | 280 | — | 100–150 |
| | | | 140 | — | 65–85 | | 200 | — | 65–85 |
| LM30 | M | | — | — | | | 150 | — | |
| | TS‡ | | — | — | | | 160 | | |
| LM28 | TE | | — | | | | 170 | | 90–130 |
| | TF | | 120 | | 100–140 | | 190 | | 100–140 |
| LM29 | TE | | 120 | | 100–140 | | 190 | | 100–140 |
| | TF | | 120 | | 100–140 | | 190 | | 100–140 |
| 2L91 | TB | 165 | 220 | 7 | | 165 | 265 | 13 | |
| 2L92 | TF | 200 | 280 | 4 | | 200 | 310 | 9 | |
| L154 | TB | 160 | 215 | 7 | | 165 | 265 | 13 | |
| L155 | TF | 200 | 280 | 4 | | 200 | 310 | 9 | |
| L119 | TF | 190 | 215 | 1 | | — | — | — | |
| 4L35 | TB | 210 | 220 | — | | 230 | 280 | — | |
| LM12 | M | — | 170 | — | | — | — | — | |
| LM5 | M | — | 140 | 3 | | — | 170 | 5 | |
| 5018A | TB | 170 | 275 | 5 | | 170 | 305 | 10 | |
| LM10 | TB | | 280 | 8 | | | 310 | 12 | |
| 4L53 | TB | 170 | 280 | 8 | | 170 | 310 | 12 | |
| DTD 5008B | Aged at room temperature | | 215 | 4 | | | | | |
| 3L51 | Artificially aged | 125 | 160 | 2 | | 140 | 200 | 3 | |
| 3L52 | TF | 245 | 280 | — | | 295 | 325 | — | |

† Average values for information only.
‡ Stress relieved by heating to an agreed temperature.

Typical mechanical properties of sand and chill castings are given in Table 26 based upon the requirements of BS 1490. These are generally in line with the recommendations of the International Standards Organisation

**Table 26** — Castings — typical mechanical properties of aluminium alloys

| Alloy designation | Temper | 0.2% proof stress N/mm² | Tensile N/mm² | Elongation % on √5.65/S₀ | Fatigue strength 50×10 cycles N/mm² | Hardness Brinell | Max service temperature °C | Modulus of elasticity N×10²N/mm² |
|---|---|---|---|---|---|---|---|---|
| | | | *(a) Sand castings* | | | | | |
| LM0 | M | 30 | 80 | 30 | — | 25 | — | 69 |
| LM16 | TB | 130 | 210 | 3 | 70 | 80 | — | 71 |
| | TF | 240 | 280 | 1 | 70 | 100 | — | 71 |
| LM4 | M | 100 | 150 | 2 | 75 | 70 | 150 | 71 |
| | TF | 250 | 280 | 1 | — | 105 | 150 | 71 |
| LM25 | M | 90 | 140 | 2.5 | — | 60 | 150 | 71 |
| | TE | 130 | 170 | 1.5 | 55 | 70 | 150 | 71 |
| | TB7 | 100 | 170 | 3 | — | 65 | 150 | 71 |
| | TF | 220 | 250 | 1 | 60 | 105 | 150 | 71 |
| LM27 | M | 90 | 150 | 2 | — | 75 | — | 71 |
| LM21 | M | 130 | 180 | 1 | — | 85 | — | 71 |
| LM6 | M | 70 | 170 | 8 | 51 | 55 | 150 | 71 |
| LM9 | TE | 120 | 180 | 2 | 55 | 70 | — | 71 |
| | TF | 220 | 250 | — | 70 | 100 | — | 71 |
| LM13 | TF | 200 | 200 | — | 85 | 115 | — | 71 |
| | TF7 | 140 | 150 | 1 | — | 75 | — | 71 |
| LM28 | TF | 120 | 130 | 0.5 | — | 120 | — | 82 |
| LM29 | TE | 120 | 130 | 0.3 | — | 120 | — | 88 |
| | TF | 120 | 130 | 0.5 | — | 120 | — | 88 |
| LM5 | M | 90 | 170 | 5 | 54 | 60 | — | 71 |
| LM10 | TB | 180 | 310 | 15 | 60 | 85 | — | 71 |
| | | | *(b) Chill castings* | | | | | |
| LM0 | M | 30 | 80 | 40 | — | 25 | — | 69 |
| LM16 | TB | 140 | 250 | 6 | 85 | 85 | — | 71 |
| | TF | 270 | 310 | 2 | 93 | 110 | — | 71 |
| LM22 | TB | 120 | 260 | 9 | — | 75 | — | 71 |
| LM4 | TF | 250 | 310 | 3 | 85 | 110 | 150 | 71 |
| LM25 | M | 90 | 180 | 5 | — | 60 | 150 | 71 |
| | TE | 150 | 220 | 2 | — | 80 | 150 | 71 |
| | TB7 | 100 | 230 | 8 | 75 | 65 | 150 | 71 |
| | TF | 240 | 310 | 3 | 95 | 105 | 150 | 71 |
| LM27 | M | 100 | 180 | 3 | — | 80 | — | 71 |
| LM21 | M | 130 | 200 | 2 | — | 90 | — | 71 |
| LM26 | TE | 180 | 230 | 1 | — | 105 | — | 71 |
| LM24 | M | 110 | 200 | 2 | — | 85 | — | 71 |
| LM6 | M | 80 | 200 | 13 | 68 | 60 | 150 | 71 |
| LM9 | M | — | 200 | 3 | — | — | — | 71 |
| | TE | 150 | 250 | 2.5 | 75 | 80 | — | 71 |
| | TF | 280 | 310 | 1 | 90 | 110 | — | 71 |
| LM20 | M | 80 | 220 | 7 | 80 | 60 | — | 71 |
| LM2 | M | 90 | 180 | 2 | — | 80 | 150 | 71 |
| LM28 | TE | 170 | 190 | 0.5 | | 120 | | 82 |
| | TF | 170 | 200 | 0.5 | | 120 | | 82 |
| LM13 | TF | 280 | 290 | 1 | 100 | 125 | — | 71 |
| | TF7 | 200 | 210 | 1 | — | 75 | — | 71 |
| LM30 | M | 150 | 150 | — | | 110 | — | 82 |
| | TS | 160 | 160 | — | | 110 | | 82 |
| LM29 | TE | 170 | 210 | 0.3 | | 120 | | 88 |
| | TF | 170 | 210 | 0.3 | | 120 | | 88 |
| LM12 | M | 150 | 180 | — | 60 | 95 | | 71 |
| LM5 | M | 90 | 230 | 10 | 100 | 65 | — | 71 |
| LM10 | TB | 180 | 360 | 20 | — | 95 | — | 71 |

(ISO) and with the values in America and the principal European countries. The members of the European Community have their own CEN requirements which are generally the same as, or very similar to, national standards. The standards for both cast and wrought materials for aircraft purposes are in separate series either as BS 'L' or as DTD Specifications.

## 5.3  WROUGHT PRODUCTS [1]

### 5.3.1  Alloy designations

Wrought aluminium and its alloys are specified in the British Standards series 1470–1475, and are classified by chemical composition in an internationally agreed system in which the first of the four digits in the designation indicates the alloy group according to the major alloying elements, as set out in Chapter 4.

*1XXX Group*

In this group, for minimum purities of 99.00 per cent or higher, the last two of the four digits indicate the minimum aluminium percentage. For example, 1060 indicates aluminium with a minimum purity of 99.60 per cent.

The second digit indicates modifications in impurity limits or the addition of alloying elements. For example, 1100 indicates aluminium of 99.00 per cent purity which includes a small addition of copper. If the second digit is zero the metal is unalloyed and contains only natural impurities within the specified limits.

*2XXX to 8XXX Groups*

In these groups the last two of the four digits have no special significance but serve to identify the different alloys in the group. The second digit indicates alloy modifications; zero indicates the original alloy.

*National variations*

When national variations of existing alloy compositions are registered they are identified by a letter after the numerical designation. The letters are allocated in alphabetical sequence, starting with 'A' for the first national variation registered but omitting 'I', 'O' and 'Q'.

Chemical compositions of wrought aluminium alloys are given in Table 27.

### 5.3.2  Temper designations

A universal system of temper designations for wrought products has not yet been accepted. However, the USA temper designation system has been accepted by the Association of European Aerospace Constructors (AECMA) for CEN Specifications.

### 5.3.2.1   *Work hardening alloys*

The non-heat treatable wrought materials whose strength is attained by alloying and cold working (strain hardening) are sometimes referred to as work hardening alloys. The degree of work hardening is expressed as a temper 'H' with a number 2, 4, 6, 8 or 9. For any one alloy, the higher the temper number, the greater the amount of cold deformation applied to the material, indicating increasing strength. The amount of cold work is usually expressed as the reduction of cross-sectional area (i.e. the cross-sectional area as a percentage of that of the soft-starter stock) which must be employed to attain a given temper, say H4 (half hard), for a given alloy and commodity, but this will not necessarily be the same as that required for a different alloy or commodity.

The fully soft, annealed condition is indicated by the letter 'O', and 'M' ('F' in the USA) indicates 'as-manufactured' material, such as hot rolled plate or extruded non-heat treatable alloy section, that has received no subsequent treatment.

The USA system differs from that of the UK only in that it inserts another figure after the 'H' to show how the temper was achieved. The figure '1' indicates that the temper was achieved by strain hardening.

The figure '2' indicates that the temper was achieved by strain hardening by more than the amount required and then subjecting the material to a low temperature (below the recrystallisation temperature) partial annealing treatment. The minimum tensile strength attained is the same as by the previous method but the elongation is slightly higher.

The figure '3' indicates that after the cold work, the properties are stabilised by a low temperature treatment. This treatment is used only for those alloys which, otherwise, would naturally age-soften at room temperature. The result of the stabilising treatment is to lower the mechanical properties slightly and to increase ductility.

The temper designation systems can be compared as follows:

| USA symbol | Description | UK equivalent |
|---|---|---|
| H | Strain hardened, non-heat treatable material | H |
| H1 | Strain hardened only | No equivalent |
| H2 | Strain hardened and partially annealed | No equivalent |
| H3 | Strain hardened and stabilised | No equivalent |
| H12, H22, H32 | Quarter hard | H2 |

| | | |
|---|---|---|
| H14, H24, H34 | Half hard | H4 |
| H16, H26, H36 | Three-quarters hard | H6 |
| H18, H28, H38 | Fully hard (hardest commercially practicable temper) | H8 |
| H19 | A special hard temper for specific applications | No equivalent |
| F | As manufactured | M |
| O | Annealed, soft | O |

### 5.3.2.2 *Heat treatable alloys*

In the USA system the prefix 'T' for 'thermal treatment' is followed by the numerals 1 to 10, indicating the ten basic heat treatment tempers. This system is capable of being extended, by the addition of figures, to take account of variations of heat treatment. The British system is not as complete. British equivalent symbols are shown in parentheses.

| USA symbol | Description | UK symbol |
|---|---|---|
| T1 | Cooled from an elevated temperature shaping process and naturally aged to a substantially stable condition | — |
| T2 | Cooled from an elevated temperature shaping process, cold worked, and naturally aged to a substantially stable condition | — |
| T3 | Solution heat treated, cold worked, and naturally aged to a substantially stable condition | (TD) |
| T4 | Solution heat treated and naturally aged to a substantially stable condition | (TB) |
| T5 | Cooled from an elevated temperature shaping process and then artificially aged | (TE) |
| T6 | Solution heat treated and then artificially aged | (TF) |
| T7 | Solution heat treated and stabilised | — |
| T8 | Solution heat treated, cold worked, and then artificially aged | (TH) |
| T9 | Solution heat treated, artificially aged and then cold worked | — |
| T10 | Cooled from an elevated temperature shaping process, cold worked and then artificially aged | — |
| T73 | Solution heat treated and artificially aged in two stages | — |
| T76 | Solution heat treated and artificially aged in two stages (different from T73) | — |

Variations on the basic systems, applied to particular alloys, can be registered with the American Aluminium Association which keeps a separate register for tempers. Some additional digits have been assigned for specific conditions, such as stress relieved tempers as follows:

T51     Stress relieved by controlled stretching. This applies directly to plate. For extrusions there are two sub-divisions:

       T510 for products that receive no further straightening after stretching

       T511 for products that may receive minor straightening after stretching to comply with tolerances

       Among the commonly used T51 conditions in British Standards are

T351   Solution heat treated, cold worked, controlled stretched and naturally aged

T651   Solution heat treated, controlled stretched, and artificially aged

T851   Solution heat treated, cold worked, controlled stretched and artificially aged

T7351 Solution heat treated, controlled stretched and artificially aged in two stages to provide optimum stress corrosion resistance, but with 10–15 per cent lower tensile properties than T6

T7651 Solution heat treated, controlled stretched and artificially aged in two stages to achieve optimum exfoliation corrosion resistance, but with about 10 per cent lower properties than T6

T52     Stressed by compressing

T54     Stress relieved by combined stretching and compressing

Specific digit designations are also assigned to indicate materials supplied in the unheat treated condition to be heat treated by the user.

T42     Test specimens solution heat treated from the O or F temper, and naturally aged to a substantially stable condition by the supplier to demonstrate response to heat treatment

T62     Test specimens solution heat treated from the O or F temper, and artificially aged by the supplier to demonstrate response to heat treatment

### 5.3.3 Wrought alloy compositions
The main alloying elements of a number of General Engineering and Aerospace Wrought Alloys are given in Table 27.

### Table 27 — Wrought alloys — chemical compositions

| Alloy designation BS 1470–1475 | Supplementary BS 1400 BS and DTD | Chemical composition wt% | | | | | | | |
|---|---|---|---|---|---|---|---|---|---|
| | | Si | Cu | Mn | Mg | Cr | Ni | Zn | Al (By difference) |
| *(a) Work hardening alloys* | | | | | | | | | |
| 1080A | | | | | | | | | 99.8 min |
| 1050A | | | | | | | | | 99.5 min |
| 1200 | | 1.00 Si+Fe | | | | | | | 99.00 min |
| 1350 | | | | | | | | | 99.50 min |
| 3103 | BS2897/8 | | | 0.9–1.5 | | | | | |
| 3105 | 3L59/60/61 | | | 0.3–0.8 | 0.2–0.8 | | | | |
| 4043A | BS 4300/6 | 4.5–6.0 | | | | | | | |
| 4047A | | 11.0–13.0 | | | | | | | |
| 5005 | BS 4300/7 | | | | 0.50–1.1 | | | | |
| 5251 | 2L80/81 | | | 0.10–0.50 | 1.7–2.4 | | | | |
| | 3L56 | | | | | | | | |
| | 4L44 | | | | | | | | |
| 5454 | BS 4300/8/11/12 | | | 0.50–1.0 | 2.4–3.0 | 0.05–0.20 | | | |
| 5554 | BS 4300/12/13 | | | 0.5–1.0 | 2.4–3.0 | 0.05–0.20 | | | |
| 5154A | | | | | 3.1–3.9 | 0.10–0.50 Mn+Cr | | | |
| 5083 | | | | 0.40–1.0 | 4.0–4.9 | 0.05–0.25 | | | |
| 5056A | 3L58 | | | 0.10–0.6 | 4.5–5.6 | Mn+Cr 0.10–0.6 | | | |
| *(b) Heat treatable alloys* | | | | | | | | | |
| | | Si | Cu | Mn | Mg | Cr | Ni | Zn | Others |
| 6463 | BS 4300/4 | 0.20–0.6 | | | 0.45–0.9 | | | | |
| 6063 | DTD 372B | 0.20–0.6 | | | 0.45–0.9 | | | | |
| 6063A | | 0.30–0.6 | | | 0.6–0.9 | | | | |
| 6101A | BS 2898 | 0.30–0.7 | | | 0.40–0.9 | | | | |
| 6061 | L117, L118 | 0.40–0.8 | 0.15–0.40 | | 0.8–1.2 | 0.04–0.35 | | | |
| 6082 | L112, L113, | 0.7–1.3 | | 0.40–1.0 | 0.6–1.2 | | | | |

| | BS 4300/5 | | | | | | | | |
|---|---|---|---|---|---|---|---|---|---|
| 2011 | | | 5.0–6.0 | | | | | | Bi 0.20–0.6<br>Pb 0.20–0.6 |
| 2031 | 2L83<br>DTD 5070B | 0.50–1.30 | 1.8–2.8 | – | 0.6–1.2 | – | 0.6–1.4 | | Fe 0.6–1.2 |
| 2618A | DTD 5084A<br>717A, 731B<br>745A | 0.15–0.25 | 1.8–2.7 | | 1.2–1.8 | | 0.8–1.4 | | Fe 0.9–1.4 |
| 2017 | DTD 150A<br>6L37<br>3L63, 2L77 | | 3.5–4.5 | 0.4–0.7 | 0.4–0.9 | | | | |
| 2014A | 2L87, 2L93<br>L103, L105<br>L156 to L159<br>L163 to L167<br>DTD 5010A<br>DTD 5030A<br>DTD 5040A | 0.50–0.9 | 3.9–5.0 | 0.40–1.2 | 0.20–0.8 | | | | |
| 2024 | 2L97, 2L98<br>L109, L110<br>DTD 5100A<br>4300/14<br>4300/15 | | 3.8–4.9 | 0.30–0.90 | 1.2–1.8 | | | | |
| 7020 | | | | 0.05–0.50 | 1.0–1.4 | 0.10–0.35 | | 4.0–5.0 | Zr 0.08–0.20<br>Zr+Ti 0.08–0.25 |
| 7010 | DTD 5120B | | 1.5–2.0 | | 2.1–2.6 | | | 5.7–6.7 | Zr 0.11–0.17 |
| | DTD 5130A | | 1.5–2.0 | | 2.2–2.7 | | | 5.7–6.7 | |
| 7075 | 2L88, 2L95<br>L160 | | 1.2–2.0 | | 2.1–2.9 | 0.10–0.25 | | 5.1–6.4 | Ti+Zr 0.20 |
| Al–Li | DTD 5110<br>DTD 5124<br>XXXA | | 1.2 | | 0.7 | | | | Zr 0.12<br>Li 2.5 |

If no minimum level is given in the specification, no reference is made to the element in the above table; however, the element may be present or necessary in the alloy.

**5.3.4  Physical properties of wrought alloys**
In general the physical properties of a given wrought alloy will vary to a
lesser extent, dependent upon form, size, shape and temper or heat
treatment condition, than do the mechanical properties. For some alloys,
physical properties have been established for different tempers or heat
treatment conditions and some of the information available is given in
Table 28 as a guide to the level of properties which may be expected. Unlike

**Table 28(a)** — Physical properties of some wrought aluminium alloys:
work-hardening alloys

| Alloy designation | Density | Coefficient of linear expansion 293–373 K $10^{-6}$/K | Thermal conductivity at 298 K W/mK | % IACS | Electrical Resistivity at 293 K μΩcm | Conductivity at 293 K % IACS |
|---|---|---|---|---|---|---|
| 99.99 % Al | 2.70 | 23.5 | 244 | 61.9 | 2.7 | 63.8 |
| 1080A | 2.70 | 24 | 230 | 58.4 | 2.8 | 61.6 |
| 1050A | 2.71 | 24 | 230 | 58.4 | 2.8 | 61.6 |
| 1200 | 2.71 | 24 | 226 | 57.4 | 2.9 | 59.5 |
| 3103 | 2.73 | 23 | 172 | 43.7 | 4.0 | 43.1 |
| 3105 | 2.70 | 24 | 172 | 43.7 | 4.0 | 43.1 |
| 5005 | 2.69 | 23.9 | 201 | 51.1 | 3.3 | 52.2 |
| 5251 | 2.69 | 24 | 155 | 39.4 | 4.7 | 36.7 |
| 5454 | 2.68 | 24 | 147 | 37.3 | 5.1 | 33.8 |
| 5154A | 2.67 | 24.5 | 138 | 35.1 | 5.4 | 31.9 |
| 5083 | 2.67 | 24.5 | 109 | 27.7 | 6.1 | 28.3 |

mechanical properties, physical properties are not usually quoted as
requirements by official specifications. An exception to this general state-
ment is the electrical resistivity or conductivity requirement for electrical
conductor and busbar materials. Another example is the French Standard
ALR 3350 which has a minimum resistivity value for the alloy AZ8GU.

Electrical conductivity, as measured by eddy-current methods, may be
used as a non-destructive test for checking the uniformity of solution heat
treatment and precipitation treatment of commodities such as plate or
extrusions.

If it is essential to obtain accurate figures for the properties of a specific
alloy in a particular form, the aluminiun producer/supplier or the relevant
specifications should be consulted.

The physical properties of General Engineering Wrought Alloys are

**Table 28(b)** — Physical properties of some wrought aluminium alloys: heat treatable alloys

| Designation | Temper | Density g/cm$^3$ | Coefficient of linear expansion 20–100 °C 10$^{-6}$/K | Thermal conductivity at 298 K | | Resistivity at 293 K μΩcm | Conductivity at 293 K % IACS |
|---|---|---|---|---|---|---|---|
| | | | | W/mK | % IACS | Electrical | Electrical |
| 6101A | TF | 2.70 | 23.5 | 214 | 54.4 | 3.133 max | 55.1 |
| 6063 | TB | 2.70 | 24 | 197 | 50.0 | 3.5 | 49.3 |
| | TF | — | 23.5 | 201 | 51.1 | 3.3 | 52.2 |
| 6082 | TB | 2.70 | 23 | 172 | 43.7 | 4.1 | 42.1 |
| | TF | — | — | 184 | 46.7 | 3.7 | 46.6 |
| 2014A | TB | 2.8 | 22 | 142 | 36.1 | 5.3 | 32.5 |
| | TF | — | — | 159 | 39.8 | 4.5 | 38.3 |
| 2024 | TD | 2.77 | 23 | — | — | 5.7 | 30 |
| | TB | — | — | 151 | 38.4 | 5.7 | 30 |
| 2031 | TB | 2.75 | 22 | 168 | 42.7 | 4.1 | 42.1 |
| | TF | — | — | 168 | 42.7 | 4.1 | 42.1 |
| 2618A | TF | 2.75 | 22 | 151 | 38.4 | 4.4 | 39.2 |
| 7020 | TB | 2.78 | — | 134 | 34.0 | 4.6 | 37.5 |
| | TF | | | 134 | 34.0 | 4.6 | 37.5 |
| 7075 | TF | 2.80 | 23.5 | 130 | 32.9 | 5.2 | 34 |

given in Table 28(a) for work hardening alloys and Table 28(b) for heat treatable alloys.

### 5.3.4.1 Density

The densities of the unalloyed aluminium series show a slight increase over 2.70 g/cm$^3$ with decrease in purity; this is mainly attributable to an increase in iron content. The manganese-bearing alloy 3103 has a density of 2.73 g/cm$^3$. In the alloy 3105, any increase due to manganese is counteracted by the magnesium content. In the 5XXX series of magnesium alloys there is a general decrease in density with increase in magnesium but here again the manganese and chromium contents modify the effect of magnesium.

The densities of the 6XXX magnesium–silicon series of alloys are similar to that of unalloyed aluminiun.

In the 2XXX copper series, the net effect of the alloying additions is to increase densities into the range 2.75 to 2.80 g/cm$^3$.

The densities of the 7XXX zinc alloys fall into the same range as the copper series, 2.75–2.80 g/cm$^3$.

### 5.3.4.2 Conductivities — thermal and electrical

*Thermal*

The thermal conductivity of unalloyed aluminium is good, being between

226 and 244 W/m°C — the higher the purity, the better the conductivity. For alloys of the 3XXX series having small contents of manganese or manganese and magnesium the value falls to 172 W/m°C. In the 5XXX series, thermal conductivity decreases with increase in magnesium content, being 201 W/m°C for 5005 reducing to 109 W/m°C for 5083.

The heat treatable 6XXX magnesium–silicon series of alloys have thermal conductivity values in the range 172 to 197 W/m°C for the TB temper and 184–214 W/m°C for the TF temper. For the TB temper, the 2XXX series have values of 151 and 152 W/m°C and for those alloys which respond to artificial ageing, values of 151–168 W/m°C are reported. The 7XXX zinc alloys have values of 130–134 W/m°C for the TF temper.

*Electrical*
Unalloyed aluminium is a very good electrical conductor — the higher the purity, the higher the conductivity. For overhead cables and busbars a purity of 99.5 per cent aluminiun is specified, typical conductivities being of the order 61.6 per cent IACS. For applications requiring good conductivity coupled with higher mechanical strength, aluminium with a higher iron content of about 1 per cent may be used. Alternatively, an alloy of the magnesium–silicon series which combines conductivity values of 55.1 per cent IACS minimum with tensile properties about 30 per cent higher than those of 1350 H8 may be used.

Manganese and magnesium alloys of the 3XXX and 5XXX series have electrical conductivities in the range 50 down to 28 per cent IACS — the higher the magnesium content, the lower the conductivity.

The conductivities of the magnesium–silicon alloys decrease with increase in alloy content. The values for material in the solution heat treated and artificially aged condition are higher than those for heat treated and naturally aged material.

In the tabulated 2XXX alloys, conductivities for the TB condition range from 30 to 42.1 per cent IACS. Artificial ageing improves the values for 2014A from 32.5 per cent in the TB condition to 38.3 per cent in the TF condition. The figures reported for 7020 and 7075 fall within the same range.

### 5.3.4.3 Elasticity
The elastic moduli of aluminium alloys are approximately the same for all the commonly used compositions and are slightly over one-third of those of steel. In some circumstances this lower value may be of considerable advantage. When an aluminium structure is loaded under shock conditions, its greater resilience enables it to absorb more energy than a corresponding steel structure. The lower value of the elastic modulus means that for equal rigidity an aluminium beam must have a greater moment of inertia than a steel beam. The geometrical shape of the section and the elastic constants,

modulus of elasticity ($E$) and modulus of rigidity or shear modulus ($G$) of the material, rather than the strength properties are the pertinent factors controlling stability and the prevention of failure due to buckling torsion and local failure [2]. For long struts or beams the stiffness can be made the same as that of a steel strut or beam provided it is possible to increase the depth of the beam or radius of gyration of the strut.

The following values are taken as approximations covering aluminium and aluminium-base alloys (except for Al–Li and high silicon alloys for which the modulus is somewhat higher).

| | |
|---|---|
| Young's modulus of elasticity ($E$) | $60.8 \times 10^3$ N/mm$^2$ |
| Torsion modulus of rigidity | |
| or shear modulus ($G$) | $24$–$26.7 \times 10^3$ N/mm$^2$ |
| Poisson's ratio | $0.32$–$0.34$ |

### 5.3.5  Mechanical properties

#### 5.3.5.1  Tensile properties

The form of the stress–strain curves for aluminium and its alloys is shown in Fig. 52, and the tensile properties normally determined are proof stress, tensile strength and elongation. Aluminium, in common with other non-ferrous metals, does not show a sharply defined yield point as does steel, and the 0.2 per cent proof stress is normally determined. BS 18 defines proof stress as 'the stress which produces, while the load is still applied, a non-proportional extension equal to a specified percentage of the extensometer gauge length'. The method of deriving proof stress is illustrated in Fig. 4 (Chapter 1).

Elongation is determined as a percentage increase of length of a stated gauge length inscribed on a tensile test piece. British Standards specify values on a gauge length of $5.65\sqrt{S_o}$ (where $S_o$ is the original cross-sectional area of the test piece). For thin material such as sheet or thin extruded section a gauge length of 50 mm may be specified.

For wrought materials the methods for test piece preparation and testing are defined in BS 4A4 'Test Pieces and Test Methods for Metallic Material' and BS 18 'Methods for Tensile Testing of Materials'.

For wrought aluminium alloys there are not really any 'typical' mechanical properties. For example, because of the differences in the amount of segregation which occurs across the diameter of large diameter ingots, used for the production of extrusions of heavy cross-sectional areas, compared with that of small diameter ingots used for small diameter bars, or sections of small cross-sectional area, the latter can be expected to give the higher tensile properties. Apart from differences in size, method of forming, type of product and nominal compositions used by suppliers, the properties of

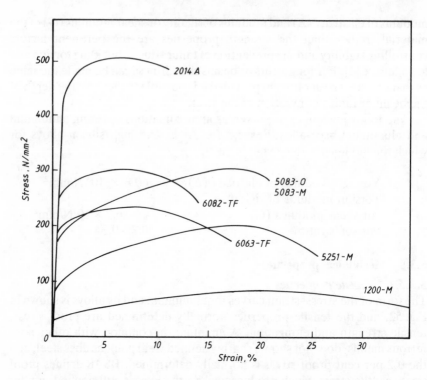

Fig. 52 — Form of stress–strain curves for typical wrought aluminium alloys.

heat treatable alloys are also affected by the rate of cooling from solution heat treatment temperature on quenching.

In Table 29 the ranges of minimum specification values are given for the tensile properties for a number of alloys and tempers, illustrating the impossibility of being able to quote a typical value for an alloy without having precise information as to the type of commodity concerned. To obtain accurate figures for a particular alloy in a particular form the aluminium producer or up-to-date editions of the relevant specifications should be consulted. Practical experience, particularly in the design of aircraft structures, has demonstrated that with careful attention to design detail and the provision of an adequate safety factor, the high proof stress to ultimate tensile ratio of the high strength structural alloys should be no detriment to their use [2].

### 5.3.5.2 Compression
For most practical purposes, proof stress in compression may be taken as the same as the proof stress in tension. It is difficult to determine the maximum

**Table 29(a)** — Specified mechanical properties of wrought aluminium alloys: work hardening alloys. Ranges of minimum values quoted for BS and DTD specifications

| Alloy designation | O temper | | | H4 temper | | | H8 temper | | |
|---|---|---|---|---|---|---|---|---|---|
| | 0.2% proof stress N/mm² min | Tensile strength N/mm² min max | Elongation On 50 mm % min | 0.2% proof stress N/mm² min | Tensile strength N/mm² min max | Elongation on 50 mm % min | 0.2% proof stress N/mm² min | Tensile strength N/mm² min | Elongation on 50 mm % min |
| 1080A | | 90 | 29–35 | | 95–120 | 5–8 | | 125 | 3–5 |
| 1050A | | 55–95 | 22–32 | | 100–135 | 4–8 | | 135 | 3–4 |
| 1200 | | 65–105 | 18–30 | | 110–140 | 3–6 | | 140 | 2–4 |
| 1350 | | 90 | 25 | | 95–140 | 8 | | 145 | 3 |
| 3103 | | 90–130 | 20–25 | | 140–175 | 3–7 | | 175 | 2–4 |
| 3105 | | 110–155 | 16–20 | 145 | 160–205 | 2–4 | 190 | 215 | 1–2 |
| 5005 | | 95–145 | 18–22 | 100 | 145–185 | 3–6 | 165 | 185 | 1–3 |
| 5251 | 60 | 160–200 | 18–20 | 175 | 225 | 5 | | | |
| 5454 | 80 | 215–285 | 12–18 | 200 | 270–325 | 3–6 | | | |
| 5154A | 85 | 215–275 | 12–18 | 200 | 245–325 | 4–6 | | | |
| 5083 | 125 | 275–350 | 12–16 | 270 | 345–405 | 4–8 | | | |

Table 29(b) — Specified mechanical properties of wrought aluminium alloys: heat treatable alloys. Ranges of minimum values quoted for BS and DTD specifications

| Alloy designation | TB or T4 temper | | | TF or T6 temper | | |
|---|---|---|---|---|---|---|
| | 0.2% proof stress $N/mm^2$ min | Tensile strength $N/mm^2$ min | Elongation on 50 mm % min | 0.2% proof stress $N/mm^2$ min | Tensile strength $N/mm^2$ min | Elongation on 50 mm % min |
| 6463 | 75 | 125 | 16 | 160 | 185 | 10 |
| 6063 | 70–100 | 120–155 | 13–18 | 130–180 | 150–200 | 6–10 |
| 6063A | 90 | 150 | 12 | 190 | 230 | 7 |
| 6101A | — | — | — | 170 | 200 | 8 |
| 6061 | 115 | 190–215 | 14 | 225–240 | 280–295 | 9–12 |
| 6082 | 100–120 | 170–200 | 13–16 | 240–270 | 280–310 | 5–8 |
| 2117 | – | 290 | — | — | — | — |
| 2011 | — | — | — | 195–225 | 295–310 | 6–8 |
| 2031 | 145–160 | 310 | 13 | 280–360 | 360–400 | 6–8 |
| 2618A | — | — | — | 260—340 | 390—430 | 5 |
| 2017 | 208 | 370 | 15 | — | — | — |
| 2014A | 215–290 | 370–400 | 7–14 | 325–390 | 400–450 | 6–13 |
| 2024 | 230–280 | 350–430 | 8–14 | — | — | — |
| 7020 | 170–190 | 280–300 | 10–12 | 270–280 | 320–340 | 8–12 |
| 7010 | | | | 410–455† | 490–525† | 4–7† |
| 7075 | | | | 480–520 | 540–580 | 4 |

† 7651 Temper.

strengths in compression of many aluminium alloys, but the design of short concentrically loaded struts may be based on the tensile figures. However, long, slender structural members in compression require special treatment [3].

### 5.3.5.3 Hardness

There is no simple relationship between the hardness and tensile strength of aluminium alloys. However, Brinell hardness numbers are specified for some casting alloys and serve the purpose of a check on uniformity within a batch of material. Several determinations must always be made to obtain a reliable result. It is desirable to maintain a predetermined ratio of $P/d^2$ (where $P$ is the applied load and $d$ is the diameter of the steel ball) and to apply and maintain the load for not less than 10–15 s. Thus, a load of 500 kg is used with a 10 mm ball and 125 kg with a 5 mm ball.

The Vickers Diamond Pyramid hardness test, involving a much smaller impression than with the Brinell ball, is used in preference on wrought alloys.

For routine inspection purposes, other methods of hardness testing such

**Table 29(c)** — Specified mechanical properties of wrought aluminium alloys: clad heat treatable alloys. Ranges of minimum values quoted for BS and DTD specifications

| Alloy designation | TB or T4 temper | | | TF or T6 temper | | |
|---|---|---|---|---|---|---|
| | 0.2% proof stress N/mm² min | Tensile strength N/mm² min | Elongation on 50 mm % min | 0.2% proof stress N/mm² min | Tensile strength N/mm² min | Elongation on 50 mm % min |
| 2618A Clad 7072 | | | | 320 | 390 | 5 |
| 2014A Clad 1050A | 230–250 | 375–390 | 13–14 | 325–380 | 400–455 | 7–9 |
| 2024 Clad 1050A | 270–275 | 405–425 | 12–15 | | | |
| 7075 Clad 7072 | | | | 420–430 | 460–490 | 8 |

as the Rockwell hardness test or one of the small portable hand instruments may be used normally for wrought materials.

### 5.3.5.4  Bearing
Bearing yield strength in riveted or bolted connections is usually taken as the stress which produces 1 per cent elongation based on the hole diameter. It is general practice to consider bearing yield stress as about 1.5 times the tensile proof stress of the material if suitable edge distances are maintained, the factor varying from 1.2 to 1.8 according to the composition of the material.

### 5.3.5.5  Shear
The average shear strength of aluminium alloys may be taken as about 60 per cent of the tensile strength, but actual values are often greater, as illustrated in Fig. 53 and Table 30.

### 5.3.5.6  Fatigue
As for other non-ferrous metals, the $S/N$ curves (Semi-range of Stress/ Number of Cycles) for aluminium alloys are not asymptotic to the $N$ axis, even after the test has been continued well beyond the number of cycles commonly applied (see Fig. 54).

In Table 27, fatigue strengths are given for castings stressed for $50 \times 10^6$

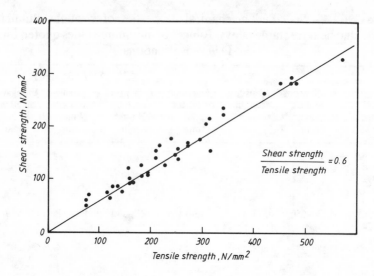

Fig. 53 — Relationship between shear strength and tensile strength of aluminium alloys.

cycles. The fatigue strengths vary in the range 50 to 100 MN/m$^2$, the average being 28 per cent of the tensile strength.

The fatigue strengths of typical wrought aluminium alloys are illustrated in Fig. 55. For a given strength the fatigue properties vary, depending upon alloy composition and temper, the ratio decreasing with increase in tensile strength. Here again the supplier should be consulted for detailed information on the fatigue properties of specific products.

In the presence of a corroding medium the fatigue life may be seriously reduced. Even in the absence of a corroding medium the fatigue life of clad material is less than that of unclad material by an amount greater than that which can be accounted for by the thickness of the cladding.

### 5.3.6   Effect of temperature on mechanical properties

#### 5.3.6.1   Properties at low temperatures

Low temperatures down to −196°C have no ill effects on the mechanical properties of aluminium and its alloys. The same applies to a few alloys which have been treated at even lower temperatures. In general the tensile strength and elongation values are both greater at sub-zero temperatures than at room temperature (see Fig. 56).

None of the alloys suffers from brittleness at low temperatures and there is no transition point below which brittle fracture occurs.

The combination of high tensile strength at low sub-zero temperatures of aluminium alloys such as 5083 (aluminium–magnesium, manganese, chro-

**Table 30** — Typical shear strength values of wrought aluminium alloys

| Alloy designation | O temper | | | H4temper | | | H8 temper | | |
|---|---|---|---|---|---|---|---|---|---|
| | Shear strength N/mm² | Tensile strength N/mm² | Ratio shear/tensile % | Shear strength N/mm² | Tensile strength N/mm² | Ratio shear/tensile % | Shear strength N/mm² | Tensile strength N/mm² | Ratio shear/tensile % |
| *(a) Work hardening alloys* | | | | | | | | | |
| 1090A | 60 | 75 | 80 | | | | | | |
| 1050A | 50 | 75 | 67 | | | | 78 | 146 | 53.5 |
| 1200 | 70 | 80 | 87.5 | 85 | 125 | 68 | 100 | 160 | 62.5 |
| 3103 | 76 | 112 | 68 | 92 | 157 | 58.5 | 108 | 195 | 55 |
| 3105 | 85 | 132 | 64.5 | 105 | 182 | 52.5 | 118 | 225 | 52.5 |
| 5005 | 76 | 120 | 63.5 | 96 | 165 | 58 | 110 | 195 | 56.5 |
| | | | | *H2 temper* | | | *H6 temper* | | |
| 5251 | 125 | 180 | 69.5 | | | | 139 | 250 | 55.6 |
| 5154A | 146 | 245 | 59.5 | 170 | 270 | 63 | | | |
| 5083 | 155 | 312 | 49.9 | | | | | | |
| *(b) Heat treatable alloys* | *TB* | | | | | | *TF* | | |
| 6063 | 131 | 155 | 84.5 | | | | 155 | 210 | 73.7 |
| 6061 | 165 | 215 | 77.0 | | | | 205 | 305 | 67.3 |
| 6082 | 178 | 225 | 79.0 | | | | 218 | 310 | 70.0 |
| 2011 | | | | | | | 235 | 340 | 69.2 |
| 2014A | 262 | 410 | 64 | | | | 293 | 470 | 62.5 |
| 2024 | 280 | 470 | 59.5 | | | | – | – | – |
| 2618A | – | – | – | | | | 280 | 450 | 62.5 |
| 7075 | – | – | – | | | | 330 | 570 | 57.8 |

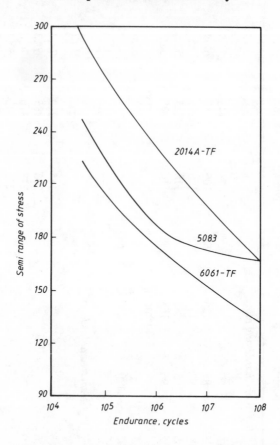

Fig. 54 — Fatigue curves for some aluminium alloys using rotating cantilever tests.
(*Courtesy Aluminium Federation.*)

mium) led to their use for the construction of large containers for the handling and storage of liquefied natural gas (LNG) before and since the introduction of special alloy steels not subject to brittle fracture.

### 5.3.6.2 Properties at elevated temperatures

Several aluminium alloys have been developed for service at elevated temperatures, the first being 'Y' alloy which had a nominal composition of 4 per cent copper, 2 per cent nickel and 1.5 per cent magnesium. It is not covered in the current BS 1490 but it is similar to BS 4L35. This heat treatable alloy is used for castings, and other alloys suitable for use at elevated temperatures include LM12 and LM13. Some typical strength-—temperature curves are given in Figs. 57 and 58.

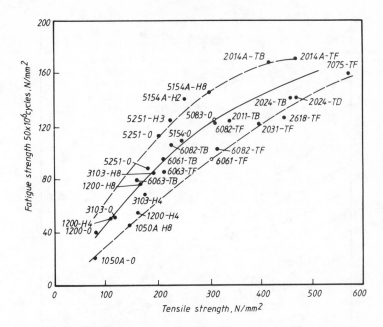

Fig. 55 — Relationships between the fatigue strength and tensile strength of some
wrought aluminium alloys.

Wrought alloys developed for use at elevated temperatures are the
general engineering alloy 2618A which is a high strength forging alloy, also
used as sheet and plate. The American alloy, AA2216, is of the same type as
the 'Y' alloy but is used for wrought products.

These materials, in addition to retaining in high degree their mechanical
properties at the higher temperature (up to about 150°C), also show good
recovery on cooling — even after prolonged periods at elevated
temperatures.

A standard testing procedure is described in BS 3688 'Methods of
Testing of Metals at Elevated Temperatures, Part 1, Tensile Testing'.

### 5.3.7  Fracture toughness

Most aluminium alloys in commercial use are not subject to unstable crack
growth in an elastic stress field; in other words, their fracture toughness is
high. For some of the higher strength alloys and tempers, particularly
aerospace materials, special tests have been developed among them the US
Navy Tear Test [17].

Investigations have demonstrated that highly significant improvements
in fracture toughness of 7075 can be obtained by reducing the volume

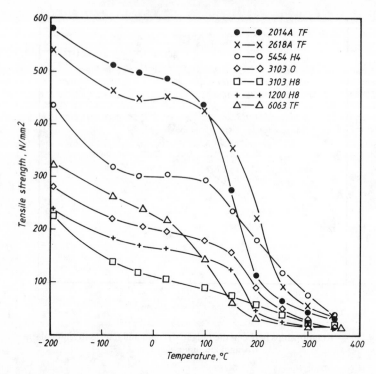

Fig. 56 — Typical tensile properties of wrought aluminium alloys at various temperatures.

fraction of insoluble (and undissolved soluble) intermetallic particles present in the alloy structure [4]. This led to the development of the alloy 7475 which differs from 7075 in the more restricted limits placed on the impurity elements, principally iron 0.12 per cent maximum and silicon 0.1 per cent maximum, which form sparsely soluble phases, and includes minor but important changes to magnesium and copper ranges. Correctly heat treated, this alloy has a toughness equal to that of 2024T3 but with higher tensile strength. Alloy 2024T3 (naturally aged temper) has been the American aircraft designers' historical preference for those structural members in which high fracture toughness is most essential [5].

In the development of aluminium–lithium-based alloys [5] having 10 per cent lower density and 10 per cent greater stiffness than current alloys, such as 2024, the major task has been to increase fracture toughness. This has been achieved with the quaternary aluminium–lithium–copper–magnesium system and by introducing cold work (by stretching) between solution heat treatment and ageing to produce a dislocation structure in the matrix to achieve optimum dispersion of the S phase ($Al_2CuMg$) and $T_1$ phase ($Al_2CuLi$).

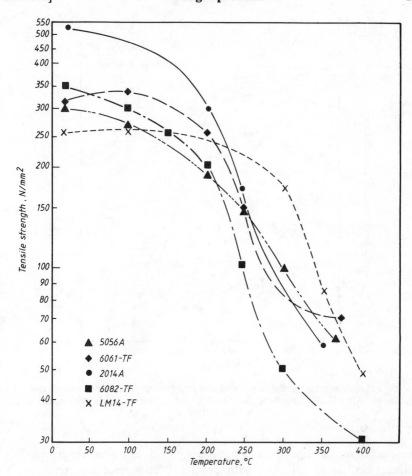

Fig. 57 — Variation of tensile strength with temperature of typical alloys soaked at temperature for 1 h prior to testing and tested at temperature. (*Courtesy Aluminium Federation.*)

### 5.3.8  Static endurance or creep strength

If a material is held under applied stress for very protracted periods at intermediate temperatures, there may be a low plastic strain, the material gradually extending: the phenomenon is usually termed 'creep'. The determination of the creep strength of aluminium and its alloys is based on the stress to produce 0.1 per cent plastic strain in either $10^3$ or $10^4$ hours.

Van Lancker [6], on the subject of the systematic study of recrystallization, recovery and polygonisation, states that studies can be made of the creep of single crystals (of various orientations), bicrystals and polycrystal-

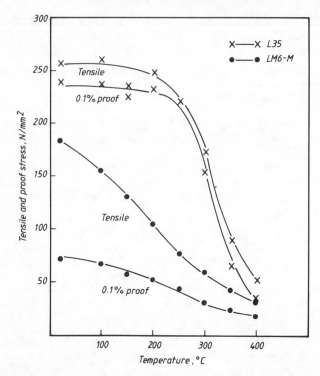

Fig. 58 — Variation of tensile strength and proof stress with temperature of two
cast alloys. Alloy L35 was soaked for 10,000 h and LM6 for 1000 h prior to testing at
temperature.

line specimens of super-pure aluminium, solid solutions containing no
obstacles and solid solutions with obstacles. It is necessary to select a
constant testing temperature, preferably characterised by the ratio of
absolute testing temperature to absolute melting point; for example, for
room temperature testing, the ratio is 293/932=0.31. If this ratio is below
about 0.4, the influence of time on plastic flow is slight.

No simple relationship exists between the tensile strength and creep
strength of an alloy.

In creep testing, strain related to time is as illustrated in Fig. 59. The
curve (at constant temperature and stress) includes three stages. (a) A
primary (or transition) stage, an initial period during which the rate of
extension falls away until it reaches a constant value in (b). (b) A secondary
(constant creep rate) stage, this steady creep stage being followed by (c), a
tertiary (accelerated creep) stage, which is a period of increasing rate of
extension. The rate of extension in stage (b) is thus the lowest, and this
minimum creep rate characterises any particular metal or alloy in a given
condition at a specific temperature and under a determined level.

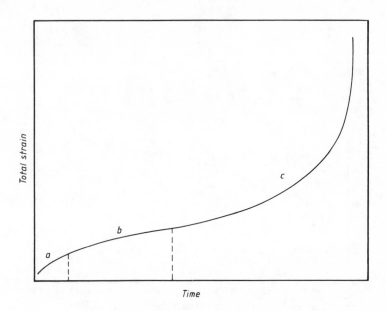

*Time*

Fig. 59 — Curve indicating general relationship between slow extension and time
under steady stress. (*Courtesy Aluminium Federation.*)

Creep is greater the higher the temperature, and while a number of aluminium alloys have useful creep performances at temperatures around 200–250°C, most of them show increasing creep at 300°C and higher temperatures. Fig. 60 shows the relationship between stress and total extension after 1000 h at 200°C for five alloys.

## 5.4 CHEMICAL PROPERTIES

Corresponding to its position in the electro-chemical series and to the high heat of formation of its oxide, aluminium is classed as a very reactive and easily oxidisable element. However, in practical use, both in the form of commercially pure aluminium and in many of its alloys, the metal possesses high resistance to the effects of weather and of many products of the chemical, food and allied industries. The high resistance of aluminium to corrosion is due to the film of oxide which begins to form immediately the metal is exposed to air and slowly increases in thickness until, after some days, no further oxidation takes place unless the film is ruptured. The normal film is hard but strongly adherent. It forms more rapidly and attains a greater thickness at high temperature. The aluminium is rendered passive by the oxide film, which is insoluble in water. Corrosion of the metal can take

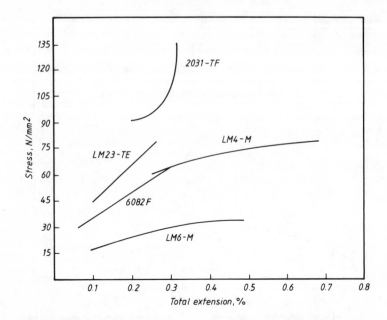

Fig. 60 — General relationships between slow extension and time under steady stress.

place only when the protective oxide film has been damaged or dissolved away.

Prolonged exposure to a moist atmosphere causes light corrosion, with a visible film which is white to grey, depending upon the alloying elements present. This light corrosion is usually easy to rub off, leaving its permanent film behind.

The protective film of oxide formed on pure aluminiun is only about 0.0000125 mm thick. The thickness of this naturally formed film can be increased and its corrosion resistance considerably improved by an electrolytic oxidation treatment known as anodising.

Certain chemical reagents rapidly dissolve aluminium, particularly hydrochloric and hydrofluoric acids, concentrated sulphuric acid, formic acid, oxalic acid and aqueous solutions of alkali hydroxides and carbonates. On the other hand, there are many reagents which have no action or very little reaction with aluminium; these include concentrated solutions of nitric acid, ammonia and most organic acids and sulphides. Aluminium alloys are usually more reactive to chemicals than pure metal but there are many exceptions to this generalisation. Aluminium–magnesium alloys, for example, are especially resistant to corrosive media containing chlorides (e.g. sea water).

Certain alkaline materials when damp, such as cement mortars, concrete and plaster, are particularly aggressive. Aluminous cements, which do not liberate lime, are less aggressive than Portland cements, which also contain sodium and potassium compounds.

Certain timbers such as oak and sweet chestnut hardwoods and Western Red Cedar and Douglas Fir softwoods which absorb moisture and liberate organic acids and soluble salts sometimes cause severe corrosion of aluminium and its alloys.

Most soils are liable to cause corrosion of aluminium and its alloys if allowed to come into direct contact with the metal. This is due mainly to the acid content of humus and also to the presence of moisture with little access of oxygen. Precautions against corrosion consist of protecting the aluminium with thick coatings of bituminous paint or inhibited paints of the zinc chromate type.

## 5.5  CORROSION

Metallic corrosion is the surface wastage that occurs when metals are exposed to reactive environments.

At temperatures where water is liquid, the predominating corrosion process is electrochemical: that is, metallic wastage occurs by anodic dissolution. Thus, even in moist air, where there is no bulk of water present, a very thin film of water may develop either by hydration of an initially formed film of oxide, chloride or other compound, or by condensation. It is ·this water that provides the solvent and connecting electrolyte needed for electrochemical corrosion. Metal first dissolves as ions, and solid products such as oxides, carbonates, sulphates and chlorides may or may not form by subsequent reaction.

At room temperature the progress of electrochemical corrosion is determined by a number of factors, foremost of which is the nature (aggressiveness or concentration) of the reactants present. The subject of corrosion is complex. Van Lancker [7] comments 'It might be said that in general corrosion testing is more or less empirical', adding that research into submicroscopic structures, alloy thermodynamics, variable voltage polarisation curves, and the physics of the solid and other products formed at the metal–electrolyte interface and allied phenomena is required for the highly complex subject to become clear.

Corrosion may be divided into eight types:

### 5.5.1  General
General corrosion is characterised by progressive and uniform thinning of the metal component. A knowledge of this rate enables 'a corrosion

allowance' to be made when designing components for service in environments of known aggressiveness.

### 5.5.2  Selective attack
Selective attack on impurities or inclusions or grain boundary constituents of specific phases of an alloy, including solid solutions, can be very harmful. Selective grain boundary attack can cause whole grains to fall out and this form of attack is especially dramatic where there is a strong rolled or extruded structure in the metal, for it produces layer corrosion (exfoliation).

### 5.5.3  Pitting attack
Pitting attack is usually very localised and the corrosion penetrates deeply in relation to the area attacked. Pitting results from variations in activity brought about by such factors as differential aeration, surface inclusions in the metal or on the surface. When it does occur, pitting can cause sudden failure in a component; this is particularly true for thin material

### 5.5.4  Bimetallic corrosion (galvanic)
Bimetallic corrosion (galvanic) can occur when two or more metals are in contact. It is characterised by the accentuated dissolution of the more reactive metal. The coupling of dissimilar metals in the presence of an electrolyte permits current to flow, giving anodic areas, the latter being corroded (see Table 31 [8]).

### 5.5.5  Crevice corrosion
Crevice corrosion is a result of differences in aeration or in ion concentration and may occur even when the components forming the crevice are of the same alloy such as lap joints, flange joints, bolt or rivet heads. Crevice corrosion can also occur when moisture condenses between the laps of coils or stacks of sheet stored under conditions of high humidity and variable temperature.

### 5.5.6  Impingement attack (erosion–corrosion)
Impingement attack (erosion–corrosion) can be encountered as the result of an otherwise protective surface film being disrupted by the impingement of entrained particles in a flowing corrodent or slurry. Another form of attack is cavitation damage caused by the collapse of low pressure air bubbles in the liquid phase.

### 5.5.7  Biological corrosion
Biological corrosion is the result of accelerated attack by microbiological organisms (bacteria) either because they manufacture aggressive species, such as proteins or sulphide ions, or because they catalyse electrochemical reactions. One strain promotes corrosion of aluminium fuel tanks in aircraft.

**Table 31** — Guide to bimetallic corrosion effects at junctions of aluminium with other metals

| Metals coupled with aluminium or aluminium alloys | Bimetallic effect |
|---|---|
| Gold, platinum, rhodium, silver<br>Copper, copper alloys, silver solder<br>Solder coatings on steel or copper<br>Nickel and nickel alloys<br>Steel, cast iron<br>Lead, tin<br>Tin/zinc plating (80/20) on steel | These metals and especially those at the top of the list are generally cathodic to aluminium and its alloys which therefore suffer preferential attack when corrosion occurs |
| Pure aluminium and its alloys not containing substantial additions of copper or zinc | When aluminium is alloyed with appreciable amounts of copper it becomes more noble, and when alloyed with appreciable amounts of zinc it becomes less noble. In marine or industrial atmospheres or when totally immersed, aluminium alloy suffers accelerated attack when in good electrical contact with another aluminium alloy which contains substantial copper such as wrought alloys 2024 and 2014 and cast alloys LM4–M and BSL92. The aluminium–zinc alloys, being less noble, are used as cladding for the protection of the stronger aluminium alloys |
| Cadmium, zinc and zinc-rich alloys | These metals are generally anodic to aluminium and suffer attack when corrosion occurs, thereby protecting the aluminium |
| Galvanised steel | The zinc coating may be preferentially removed to expose the bare steel |
| Magnesium and magnesium-based alloys | Attack on magnesium is accelerated in severe environments such as marine and industrial or under conditions of total immersion. Attack on aluminium may also be accelerated |
| Titanium<br>Stainless steel (18/8, 18/8/2) and 13% Cr)<br>Chromium plate | These metals form protective films that tend to reduce bimetallic effects. Where attack occurs the aluminium base material suffers |

**5.5.8   Stress corrosion**
Stress corrosion results from the combined action of mechanical stress and corrosion, producing an accelerated effect. This may be thought of as a special form of crevice attack since the cracks that develop form a self-perpetuating region of localised attack. The stress may be externally applied during service, or it may be internal as a result of non-uniform cold working or cooling such as quenching from solution heat treatment temperature, or due to deformation imposed during assembly.


**5.6   CORROSION CHARACTERISTICS OF ALLOYS**

Information specific to the principal alloy groups is summarised in Table 32 and in the sections which follow.


**5.6.1   Aluminium of 99.00 per cent purity and higher**
Aluminium has good resistance to atmospheric corrosion in rural, industrial and marine environments — the higher the purity, the greater the resistance to corrosion. Nodular pitting, a form of severe pitting which causes failure of aluminium water service pipes, has been investigated by the British Non-Ferrous Metals Association, who ascertained that for this type of attack to occur the water must contain calcium bicarbonate, chlorides, copper salts and dissolved oxygen. As little as 0.02 ppm of copper is sufficient to initiate severe attack [9].


**5.6.2   Aluminium–copper alloys**
Of the normal alloying elements, copper in particular effects a strong reduction of the corrosion resistance of aluminium. Nickel has a similar effect and to a lesser extent so does iron. The unclad aluminium–copper series of alloys is susceptible to pitting and intercrystalline corrosion when exposed to corrosive media such as a marine atmosphere. Material in the annealed condition is most susceptible to attack, correctly heat treated material giving the best resistance. Sheet and other wrought forms of alloys, such as 2014A type, have given good, long service without serious corrosion of the intercrystalline type. The pitting type of attack in these cases is, however, at first more severe than for Al–Mg–Si-type alloys, but is still self-stopping. It has been suggested that [10] provided the thickness of metal does not fall below a certain minimum, dependent on the alloy and its corrosion resistance in the particular environment under consideration, the effect of corrosion on the proof stress or design stress will be negligible. The limiting thickness for a heat treatable alloy of the 2014A type is given as

**Table 32** — Resistance to atmospheric attack and suitability for anodising, plating and vitreous enamelling — wrought alloys

| Material designation | Resistance to atmospheric attack | Suitability for anodising | | | | Plating | Vitreous enamelling |
|---|---|---|---|---|---|---|---|
| | | Protective | Colour | Bright | Hard | | |
| *(a) Work hardening alloys* | | | | | | | |
| 1080A | E | E | E | V–E | E | — | — |
| 1050A | V | E | E | V | E | V | G |
| 1200 | V | V | V | G | E | V | G |
| 3103 | V | G | G | P–F | G | G | V |
| 3105 | V | G | G | P–F | G | — | — |
| 5005 | V | E | E | E | E | — | — |
| 5251 | V | V | V | G–V | E | X | U |
| 5454 | V–G | V | V | G | E | — | U |
| 5154A | V–G | V | V | G | E | X | U |
| 5083 | V–F | V | V | G | E | X | U |
| *(b) Heat treatable alloys* | | | | | | | |
| 6063 | V–G | V | V | G–V | E | X | Y |
| 6061 | V–G | G | G | F | V | X | Y |
| 6082 | V–G | G | G | F | G–V | X | Y |
| 2011 | P | F | F(D) | U | G | — | U |
| 2014A | P | F | F(D) | U | G | V | U |
| 2024 | P | F | F(D) | U | G | V | U |
| 2618A | F | F | F | U | F | — | — |
| 7020 | G | F | F | † | G | — | — |
| 7075 | F | F | F | † | F | — | — |

E = Excellent. V = Very good. G = Good. F = Fair. D = Dark colours only. U = Unsuitable. P = Poor. X = A special treatment is required for successful plating. Y = A special treatment is required for successful vitreous enamelling. † = Variable response, depending on actual composition and heat treatment.

2.75 mm (0.110 inch) for severe conditions of exposure in a marine atmosphere and after allowing a factor of safety of two.

The corrosion resistance properties of the free machining alloy 2011 are inferior to those of 2014A and 2024 owing to the adverse effect of the lead and bismuth constitutents which are added to impart good chip formation on machining.

The corrosion resistance of the 2XXX alloys can be improved by anodising the surfaces, and/or by applying a protective coating such as a zinc chromate primer followed by normal painting. For rolled sheet and strip material, cladding is the normal means of protecting the core alloy. The cladding material must have a higher intrinsic corrosion resistance than the core and it may also protect sacrificially any parts of the strong alloy core which may become exposed to corrosive environments at damaged areas, even at cut edges. The first commercial clad sheet produced in 1928 was a sandwich of 2014-type alloy, having pure aluminium on each side, with a thickness equal to 5 per cent of the total sheet thickness. Other combinations were developed subsequently, e.g. an aluminium/1–2 per cent zinc alloy to protect the high strength 7XXX-type alloys. The cladding of 7075 sheet with the heat treatable 7008 Al–Zn–Mg-type alloy gives strength advantages coupled with higher resistance to abrasion and fatigue [1].

The cladding is bonded metallurgically to the core during manufacture and must, of course, be anodic relative to the core. Its composition should not change as a result of reheating or heat treatment at such a temperature as would cause migration of alloying elements from the core or reheating or heat treatment for a prolonged time at the correct temperature. Fig. 61 shows photomicrographs of an example of a 2024 alloy sheet clad with aluminium, as rolled in (a) and after re-heating for a prolonged time during which the effect of thickness of the cladding was reduced by migration, as illustrated in (b).

### 5.6.3  Aluminium–manganese alloys

Sheet material of the 3103 type has been widely used as roofings and sidings material over a great number of years, and on exposure to marine atmosphere this series has better corrosion resistance than the commercially pure grades of metal. Tests have been conducted by ALCOA over a period of years, exposing 3003 alloy (this differs from 3103 in that it has a small copper addition) to coastal, industrial and tropical environments. The pitting corrosion rates of 3003 sheet 0.36 inch (0.9 mm) thick were:

| | |
|---|---|
| Point Judith (Coastal) | 0.00011 in/yr (0.00275 mm//yr) |
| New Kensington (Industrial) | 0.00008 in/yr (0.00203 mm/yr) |
| Georgetown (Tropical) | 0.00003 in/yr (0.00075 mm/yr) |

Initially the rates were much higher than this but there was a considerable slowing down of corrosion penetration after the first year, the pitting rates

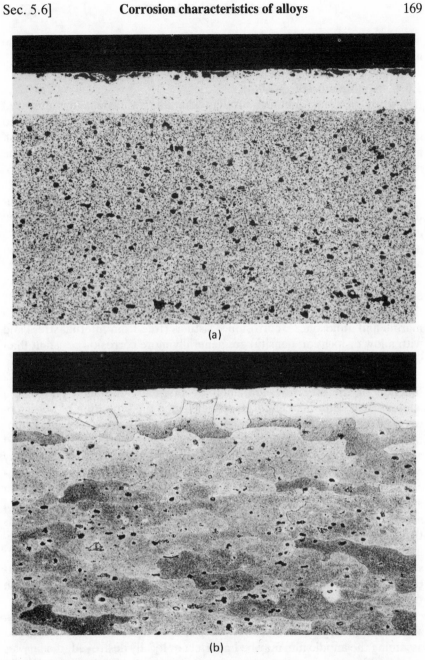

(a)

(b)

Fig. 61 — (a) Clad AA2024 alloy (clad with AA1200 alloy), as rolled. (b) Clad AA2024 alloy (clad with AA1200 alloy) after heat treatment, showing the annealed grain structure of the cladding and evidence of diffusion of copper from the core material into the clad layer. Both etched with Keller's reagent. ×100.

becoming constant after the second year. The figures quoted represent the maximum rates of pitting, average rates being much less.

### 5.6.4 Aluminium–silicon alloys

Relatively small amounts of wrought aluminium–silicon alloys are used nowadays for applications other than as brazing alloys. They have been used in the past for architectural applications where the dark grey anodised finish has proved attractive. These alloys are also used for welding rods so that the resulting weld metal behaves in much the same manner as a chill casting of the same composition (e.g. BS LM6).

Many of the general engineering casting alloys have high silicon contents, LM6 and LM18 having 12 per cent and 5 per cent silicon respectively, and possess good corrosion resistance, whereas LM4–3 per cent copper, 5 per cent silicon, 0.5 per cent manganese is rated as having medium corrosion resistance.

### 5.6.5 Aluminium–magnesium alloys

Certain aluminium–magnesium alloys are the best available for resistance to marine and atmospheric corrosion. Castings (free from internal stress and with a low dislocation density) are generally more corrosion resistant than wrought alloys.

The general corrosion resistance of aluminium–magnesium alloys is a function of the structures obtained by thermal treatment. The silicon content is an additional factor dependent upon the formation of $Mg_2Si$. In wrought alloys the corrosion phenomena depend primarily on the distribution of zones and GP compounds preceding the formation of $Al_3Mg_2$ and $Mg_2Si$. If the magnesium in an aluminium–magnesium alloy undergoes any changes in bonding with time and temperature by diffusing towards vacancies or dislocations and forming zones or GP compounds (together with adjacent magnesium denuded and thus less negative areas) the distribution at the surface will be similarly modified and so will be the reactivity towards the corrosive medium. Both intergranular and stress corrosion susceptibility increase with increases in magnesium content and service temperature; stabilisation or stress relieving of alloys having magnesium contents at temperatures around 200°C may sensitise them to intercrystalline corrosion through the formation of a continuous network of $\beta$ ($Al_3Mg_2$).

If the aim is to produce a wrought Al–Mg alloy of maximum resistance in sea water or an aggressive natural water of known pH without anodising or assuming the anodic film may be imperfect or locally destroyed, the answer must lie in finding the correct heat treatment to produce substitutional solid solution that will remain stable over long periods [11]. Between 280 and 450°C there is a range in which annealing is possible, and is accompanied by the solution of an equilibrium, $Al_3Mg_2$.

### 5.6.6   Aluminium–magnesium–silicon alloys

The medium strength heat treatable alloys of the aluminium–magnesium–silicon type offer the best compromise to sacrifice in strength to obtain immunity to intercrystalline type of attack required of materials used in buildings. These alloys are satisfactory in urban exposures but are attacked more rapidly than aluminium–magnesium alloys, though they are less susceptible to stress corrosion.

### 5.6.7   Aluminium–zinc alloys

7075-T6-type alloy in large, thick sections aged isothermally for twenty-four hours at 120°C or in two stages at temperatures of less than 150°C has inadequate resistance to stress corrosion cracking. Precipitation at temperatures of 160°C or higher progressively improves stress corrosion resistance. Higher strength combined with stress corrosion resistance is obtained when the higher temperature is preceded by a lower stage to nucleate a finer, higher density precipitate dispersion — 7075-T73 [12].† Theory [13,14] attributes stress corrosion cracking (s.c.c.) in T6 temper to selective dissolution of precipitate free zones (p.f.z.) accelerated by tension stress, and the higher resistance of T73 temper to equalisation of the electrochemical potentials. The principal microstructural difference between the T6 and T73 tempers, however, is in the intergranular precipitate which changes in structure from predominantly coherent g.p. zones of T6 to the metastable $\eta'_{4+}$ of T73. The change in precipitate structure is the basis of the dislocation–matrix precipitate interaction hypothesis [15, 16].

7075-T6 is susceptible to exfoliation corrosion under some conditions whereas T73 provides high resistance to exfoliation but incurs a strength reduction of about 15 per cent. The alloy 7050 has higher zinc and copper contents relative to 7075 zirconium in place of chromium, and more restrictive limits on the major elements as well as impurity elements. Zirconium, functioning as the high temperature precipitate forming addition to suppress recrystallization, reduces strength loss with decreasing quench cooling rate characteristic of the chromium and manganese-containing alloys.

### REFERENCES

[1] *The Properties of Aluminium and its Alloys*, Aluminium Federation, p. 6.

[2] Brimelow, E. I., *Aluminium in Building*, Macdonald, London, p. 195.

---

† Equilibrium phase field extends across the quaternary diagram joining $MgZn_2$ and $AlCuMg$. Evidence from electrochemical solution potential changes indicates Cu precipitates in the transition form of this phase during the higher temperature stage of the T73 treatment.

[3] Brimelow, E. I., *ibid.*, p. 137.

[4] Hunsicker, H. Y., Development of Al–Zn–Mg–Cu alloys for aircraft, *Phil. Trans. R. Soc., London,* **282**, 374.

[5] Grimes, R., Cornish, A. J., Miller, W. S. and Reynolds, M. A., Aluminium lithium based alloys for aerospace applications, *Metals and Materials,* June 1985, pp. 357–363.

[6] Van Lancker, M., *Metallurgy of Aluminium Alloys,* 1967, Chapman & Hall, London, p. 78.

[7] Van Lancker, M., *ibid.*, p. 311.

[8] Table 34 Taken from 'The Properties of Aluminium and its Alloys', *Guide to Bimetallic Corrosion Effects at Junctions of Aluminium and Other Metals,* Aluminium Federation.

[9] Porter, F. C. and Hodden, S. E., Nodular pitting of aluminium alloys, *J. Appl. Chem.,* 1953, **3**, 385–409.

[10] Champion, F. A., Metal thickness and corrosion effects, *Metal Industry,* 7 Jan 1949, **74**(1), 1.

[11] Van Lancker, M., Metallurgy of Aluminium Alloys, 1967, Chapman & Hall, London, p. 391.

[12] Hunsicker, H. Y., Development of Al–Zn–Mg–Cu alloys for aircraft, *Phil. Trans. R. Soc., London,* **282**, 367.

[13] Sprowls, D. O. and Brown, R. H., Stress corrosion mechanisms for aluminium alloys, *Proceedings of International Conference on Fundamental Aspects of Stress Corrosion Cracking,* NACE 466–506, 1969.

[14] Lifka, B. W. and Sprowls, D. O., Significance of intergranular corrosion in high strength aluminium alloy products, *ASTM*, STP 516, pp. 120–144, 1972.

[15] Spiedal, M. Q., Current understanding of stress corrosion crack growth in aluminium alloys, *The Theory of Stress Corrosion Cracking in Alloys,* Brussels, 1971, NATO Scientific Affairs Div. Pub., pp. 289–344.

[16] Speidel, M. O. and Hyatt, M. V., Stress corrosion cracking of high strength aluminium alloys, *Adv. Corrosion Sci. Technol., Plenum Press, 1972,* **2**, 115–335.

[17] *The Properties of Aluminium and its Alloys,* Aluminium Federation, p. 53.

**FURTHER READING**

*The Properties of Aluminium and its Alloys,* Aluminium Federation.

*The Properties and Characteristics of Aluminium Casting Alloys,* Association of Light Alloy Refiners.

*Aluminium in Building,* E. I. Brimelow, Macdonald, London.

*Basic Corrosion and Oxidation,* J. M. West, Ellis Horwood Ltd.

Effects of alloying elements on corrosion resistance of casting alloys, D. C. G. Lees, *Light Metals,* **14**, 162 (Sept. 1951).

Atmospheric corrosion and stress corrosion of aluminium copper magnesium and aluminium magnesium silicon alloys on the fully heat-treated condition, G. J. Metcalfe, *J. Inst. Metals*, 1952–53, **81**, 289.

Aluminium as a structural material, M. Bridgewater, *Engineering* 1973, **175**, 762 and 793.

*Symposium on Atmospheric Exposure Tests on Non-ferrous Metals*, E. H. Dise, R. R. Mears, Feb. 27 1946. Published by the American Society for Testing Materials.

Aluminium and Aluminium Alloys in the Food Industry with Special Reference to Corrosion, and its Prevention, J. M. Bryan, Department of Scientific and Industrial Research, *Food Investigation Special Report No. 50,* HMSO, London, 1948.

# 6

# Production of semi-fabricated forms

Aluminium and its alloys, as specified in British Standards, DTD Aerospace specifications and standards of other nationalities, together with those covering individual customer requirements, offer a wide choice to users in every sphere of activity. The diversity of alloy compositions and forms of commodity are such that it is only possible to deal with them in broad outline in this chapter.

Some half a century ago it was not uncommon for individual plants in the UK to cover a whole range of processes such as the melting and casting of ingot for hot rolling or extrusion and the fabrication of plate, sheet, extrusions, tube and wire. In the recent past there has been a change towards specialisation in the fabrication of one class of commodity.

Reduction plants are located near to sources of abundant power supply, often in remote locations, but in contrast, plants supplying semi-fabricated forms are located in areas more suited for the easy distribution of their products. These semi-fabrication plants are often at some considerable distance from the reduction plants supplying their raw material in the form of ingot for remelting and alloying. These ingots are of various grades of pure aluminium or of master alloys such as aluminium 25 or 50 per cent copper, aluminium 7 or 12 per cent manganese, aluminium 12 or 20 per cent silicon, in a form suitable for making alloys for the production of castings or wrought products.

With the development of casting techniques such as the Direct Chill (DC) process and the various continuous casting processes referred to later in this chapter it is now possible for rectangular ingots for hot rolling and coils of continuously cast reroll stock and other commodities to be supplied from the reduction plants to the semi-fabricators, saving one remelting operation and reducing the amount of manual labour required. The rolling and extrusion plants still have to dispose of process scrap, so there still remains the necessity for an integrated melting and casting facility, at the plants or within easy access.

Foundries making aluminium and aluminium alloy castings range from large integrated units down to small jobbing foundries. Their raw materials consist of ingot for remelting in the form of pure aluminium and master alloys, ingot of the correct nominal composition which may be supplied direct from the reduction plant or from secondary melters and made from reclaimed metal, and their own process off-cuts and rejects.

In the following sections the basic principles applied to the production of semi-fabricated forms are outlined, as shown in Fig. 62.


## 6.1 MELTING

The types of furnace used for the melting of aluminium and its alloys range from small crucible furnaces to large reverberatory furnaces of 25 tonne capacity, the former being charged manually with the larger furnaces designed for mechanical charging, stirring and cleaning. The method of heating may be gas, fuel oil or electricity. For the intermediate range of furnaces, electrical heating may be by radiants in the roof or by low frequency induction. Solid fuels such as coke or anthracite have been used for reverberatory furnaces but for preference, gas, electrical radiants or fuel oil are used for large furnaces. The choice is governed by considerations of availability and cost of fuel, ease of automating control, fuel efficiency, man-power requirements and maintenance costs. High sulphur-bearing fuels are not considered to be suitable for the melting of aluminium if sodium-bearing fluxes are being used, since there is a possibility of sodium sulphate being formed and this compound in contact with molten aluminium can react explosively.

The selection of refractories for the lining of aluminium melting furnaces is very critical since reactions between the molten metal and constituents such as silicon, iron or chromium in the refractory can lead to detrimental contamination of the metal being melted. Fire clay refractories containing from 47 to 99 per cent alumina continue to be used because of the availability of suitable clays in most parts of the world.

Phosphate-bonded high-alumina bricks (about 88 per cent $Al_2O_3$ and 8 per cent $SiO_2$) heated during manufacture to a temperature above the required operating temperature, when set in the appropriate phosphate-bonded mortars, can be expected to outlast other refractories commonly used for furnaces handling molten aluminium. Magnesite refractories are not readily attacked by molten aluminium; they have good abrasion resistenace and have low porosity but will not withstand thermal shock well.

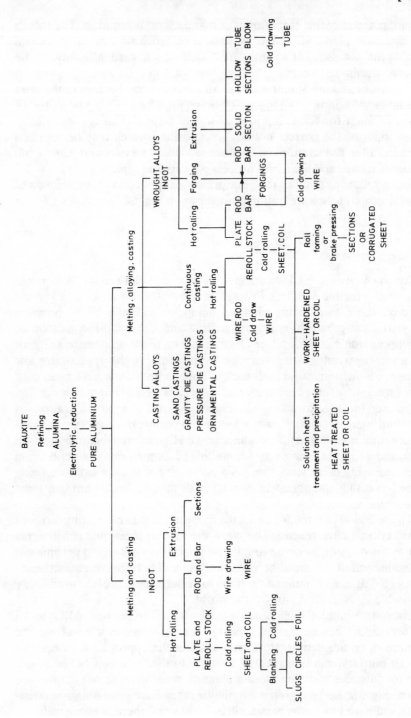

Fig. 62 — Semi-fabricated products and methods of manufacture.

It is common practice to perform melting and alloying in large furnaces
maintained at a constant temperature to give fast melting rates. A heel of
molten metal being maintained in the furnace after each transfer cycle
makes it possible for light scrap to be submerged for melting, thus reducing
oxidation losses. At intervals, metal from the melting furnace is transferred
to a holding furnace of smaller capacity, the two furnaces being at different
levels to facilitate the transfer which is by launder, siphon or tilting of the
melter. Cascading of metal during transfer must be kept to a minimum to
reduce oxide formation and entrapment.

The melting of very light material such as foil or machine swarf and
shavings is preferably performed in low frequency induction furnaces of the
coreless type. In channel-type induction furnaces, oxide build-up in the
channels necessitates frequent cleaning. Melting in induction furnaces
avoids exposing the finely divided charge to the higher temperatures
employed in reverberatory furnaces.

## 6.2  ALLOYING

The low melting point alloying elements, zinc and magnesium, may be
added direct to a furnace charge of molten aluminium held at about $700 \pm$
$10°C$. The zinc, being denser than aluminium, sinks to the floor of the
furnace and gentle stirring is required to ensure even distribution throughout the charge. At the same time, care must be taken to avoid damage to the
refractory and distribution of non-metallic inclusions in the metal. Magnesium, being less dense than aluminium floats on the surface of the melt and
burns away unless precautions are taken to plunge the bars or ingots below
the surface. This can be done by placing the magnesium in a perforated mild
steel container which is suitably coated with a refractory wash. The container must be moved around in the molten aluminium to distribute the
magnesium as it melts.

The high melting point metals, copper and nickel, react exothermically
with molten aluminium and are seldom introduced direct into melts for
castings or wrought products. Additions of these metals are normally made
in the form of master alloys such as aluminium–25 per cent copper,
aluminium–50 per cent copper, or aluminium–20 per cent nickel. Other high
melting point metals, such as manganese and silicon, are also added to melts
in the form of ingots of master alloys such as aluminium–12 per cent
manganese or aluminium–12 per cent silicon.

As an alternative to using master alloys, a method has been developed
for the introduction of alloying elements into the melt in the form of a
metallic powder, an inert gas being used to inject the powder into the melt
and to give a stirring action. This method offers a means of making minor
corrections to melt compositions, the large surface area of the powder

promoting rapid dissolution and the inert gas serving to give uniform distribution of the alloying element.

## 6.3  GRAIN REFINING

Grain refining elements such as titanium, titanium–boron, zirconium, etc., are added to the melt in the form of master alloys. These additions are introduced into the melt at as late a stage as possible to reduce the chances of the nucleating constituents coagulating and settling to the bottom of the furnace. The nucleants may be added to the molten metal by rod injection into the metal stream in the launder (refractory-lined trough) between the furnace and the casting unit. Nucleants may also be added to melts as chemical fluxes which react with the molten metal to give both a grain refining and a degassing action.

## 6.4  SAMPLING FOR ANALYSIS

To ensure that material complies with specification limits, samples are taken from the bath of metal before casting is commenced. Rapid analyses by Quantometer allow any necessary corrections to be made. Additional samples may be taken at intervals during the casting cycle. A company's internal limits for the main alloying elements are usually restricted to a narrower range than those permitted by national and international standards. In this way it is possible to adopt standard processing treatments designed to give a consistent product quality.

## 6.5  DEGASSING AND CLEANING

The melting of aluminium and its alloys yields metal containing dissolved hydrogen and solid impurities (oxides, nitrides, spinels, etc.). The amount of gas in solution increases with increase in metal temperature. The gas and solid impurities retained on solidification are a source of defects in finished products and should be eliminated before casting ingots for wrought products. However, for some complicated types of sand casting and gravity die casting it is useful to retain a small gas content, since this promotes uniformity of the material within the casting.

A number of processes are available for the reduction of the hydrogen content of molten aluminium and its alloys. Preferably the metal should not be heated to a temperature exceeding 780°C. The molten metal should be treated by bubbling an inert gas such as argon, a mixture of an inert gas and chlorine, or chlorine for degasification. Alternatively, hexachlorethane ($C_2Cl_6$) may be plunged into the melt, where it decomposes to liberate chlorine in an active state to remove hydrogen.

The finer the dispersion of the gas bubbles and the greater their distance of travel through the metal (the longer their dwell time), the greater the efficiency of the degassing action. In addition to the reduction in the hydrogen content the bubbles of gas also remove some of the suspended non-metallics. Further cleansing can be achieved by filtration of the metal through fine mesh glass cloth, rigid porous refractory or graded beds of refractory.

Untreated metal may have about 1 ml hydrogen/100 g metal retained in supersaturated solid solution on freezing. Porosity occurs as soon as a threshold value is reached. Threshold values depend upon alloy composition and are of the order 0.1–0.15 ml $H_2$/100 g Al [1].

To get gas contents down to below 0.1 ml $H_2$/100 g Al by vacuum, the pressure would have to be reduced to below 0.5 mm Hg, but this would give rise to loss of magnesium and zinc.

## 6.6   CASTINGS [2]

### 6.6.1   Sand casting
*Sand Casting* is the least costly method when relatively few castings are required and one that can be applied to almost any type of aluminium casting whether it be small or large, simple or complex. When necessary, it is possible to achieve a smoothness of surface approaching that associated with die castings. There are few technical limits to casting by this method though it is not easy to cast sections of adequate strength below about 5 mm thickness. For sand castings, the riser is larger for aluminium alloys than for heavy metals to ensure good feeding of the mould as cooling and shrinkage take place (see Fig. 63).

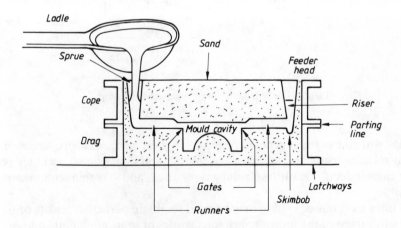

Fig. 63 — Schematic section of sand casting in its flask or moulding base.

### 6.6.2  Gravity die casting

The choice of *Gravity Die Casting* (known in North America as Permanent Mould) as a method of production is determined by (a) the number of identical castings required — usually the cost of die making is justified for runs of between 1000 and 5000 — and (b) the mechanical properties required in the function the castings will perform. Casting in metal dies tends to promote small grain formation which, in the case of aluminium alloys, is associated with higher mechanical properties. The method, although very versatile, imposes some practical limitations when castings (and consequently dies and machines) are very large and problems associated with feeding thin sections and with stresses due to shrinkage are accentuated.

### 6.6.3  Pressure die casting

The cost of making dies for *Pressure Die Casting* (termed in North America 'Die Casting') is usually high and in general can be recovered only from a production run of ten thousand or more castings; but it can be economical for smaller quantities if castings are small and of very simple design (see Fig. 64).

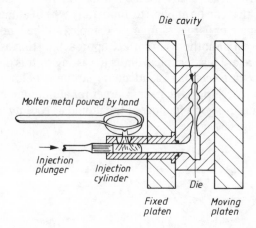

Fig. 64 — Schematic section of pressure die casting.

Pressure die castings generally have the highest mechanical properties of all in relation to weight; surfaces are smooth and dimensional accuracy is very close indeed. Very little finishing is required, and very thin sections are possible.

Other methods of casting may be used to obtain particular results or to take advantage of the individual characteristics of some aluminium alloys.

### 6.6.4   Low pressure die casting

*Low Pressure Die Casting* makes use of compressed air at low pressure to force metal up a vertical duct into a die positioned at its head. When the die is filled and the metal solidified, the pressure is relaxed and the unused metal flows back to the holding bath. The method is economical in heat, with little wastage of metal. The castings are sound, accurate and of very good appearance.

### 6.6.5   Suction casting

With *Suction Casting*, metal is drawn into the die by withdrawing air from it. The method is very effective in the production of difficult castings such as deeply finned cylinder heads used in aircraft and other engines.

### 6.6.6   Investiment casting

*Investment Casting* is the method used for casting shapes so complex that they could not be extracted from moulds or dies of the type used in the main processes. The plaster model (or pattern) is covered with warm wax to the thickness required in the metal casting. When set, this is coated, or 'invested', with a layer of plaster which is allowed to harden. The whole is then heated until all the wax runs away. The cavity remaining is filled with molten metal and when this has solidified all the plaster is removed, from within and without, leaving a faithful reproduction of the pattern. This method is also known as the 'cire-perdu' or 'lost wax' process.

The method can be used to give greater dimensional accuracy. In industry, a master pattern of steel may be used to produce soft metal, split dies, around which a wax pattern is formed. The wax pattern is 'invested' as above and casting proceeds. Because the dies are in sections they can be removed and used again.

### 6.6.7   Shell moulding

As the name implies *Shell Moulding* involves a 'shell' which is sand bonded with a thermo-setting resin. A metal pattern plate is first heated and then inverted in a 'dump box' where it is sprinkled with the sand. This is left for a time to allow the sand/resin mixture to adhere to the pattern, and is then inverted to allow the surplus sand to fall away. The assembly is cured by baking for a short time at a moderate temperature until a strong, thin shell is obtained which is stripped from the pattern.

The process is normally used for castings which are not too intricate in design. It can be operated with simple equipment, but automatic mechanized units give high rates of productions. Castings are to close tolerances and have good surface appearance. Moulds may also be made of special refractory plasters. These are baked to a hardness which allows up to about 200 castings to be made from the same mould.

The aluminium–silicon series of alloys is easily cast, being suitable for thin sections where free flow of metal is especially important.

The 5 per cent silicon, 3 per cent copper LM4 casting alloy is readily cast by all methods, has satisfactory strength, good machinability and pressure tightness and a resistance to corrosion which is adequate for many applications and may be regarded as a general-purpose material.

Information on the heat treatment, quench media and precipitation treatment for casting alloys is given in Table 33 [3].

## 6.7 CASTING FOR WROUGHT PRODUCTS

### 6.7.1 Direct chill casting

The 'semi-continuous' direct chill (DC) casting process is most frequently used for the production of rectangular ingots for rolling to plate, sheet and foil, and cylindrical ingots for extruded rods, bars, sections, hollow sections, tube and wire rod. Ingots of square section are cast for rolling to bars and wire rod (see Fig. 65).

The DC ring mould is made of aluminium alloy or copper-base alloy having good thermal conductivity. Moulds are typically 100 to 150 mm deep and the thin walls of the mould are cooled by a constant supply of cold water. The base of the mould can be lowered, mechanically or hydraulically, at a speed which can be set to suit the alloy and size of ingots being cast. Lowering speeds for large ingots may be of the order of 35 mm/min and for small-diameter ingots 150 to 200 mm/min.

At the commencement of casting, molten metal is poured into the mould at a steady rate. On contact with the base and water cooled skirt of the mould, a solidified cup or shallow dish shaped shell of metal is formed. As the metal level approaches to within a short distance from the top of the mould, the lowering mechanism supporting the base is set in motion at the predetermined speed, and the pouring rate is adjusted to maintain a constant level of metal in the mould. The solidified shell holding a molten core is subjected to direct chilling from water sprays directed onto the emerging ingot. Pouring continues until the desired ingot length has been cast. The length is limited by the stroke of the lowering device, the furnace capacity and the number of ingots being cast simultaneously. For small-diameter ingots, as many as eighty, in lengths of 3 to 4 m, may be cast in one 'drop' (one stroke of the lowering device), automatic devices being employed to maintain a steady feed to each individual mould. One method is to equip each mould with a refractory hot top and to feed all of these from a common header.

**Table 33** — Heat treatment data for casting alloys [3]

| Material designation and temper | Solution treatment Temperature °C | Time† h | Quench‡ medium | Precipitation treatment Temperature °C | Time§ h |
|---|---|---|---|---|---|
| **BS 1490** | | | | | |
| LM4-TF | 505–520 | 6–16 | Hot water | 150–170 | 6–18 |
| LM9-FE | — | — | — | 150–170 | 16 |
| -TF | 520–535 | 2–8 | Water | 150–170 | 16 |
| LM10-TB | 425–435 | 8 | Oil at 160°C max | | |
| LM13-TE | | | | 160–180 | 4 16‖ |
| -TF | 515–525 | 8 | Hot water | 160–180 | 4 16 |
| -TF7 | 515–525 | 8 | Hot water | 200–250 | 4 16 |
| LM16-TB | 520–530 | 12 | Hot water | — | — |
| -TF | 520–530 | 12 | Hot water | 160–170 | 8–10 |
| LM22-TB | 515–530 | 6–9 | Hot water | — | — |
| LM25-TB7 | 525–545 | 4–12 | Hot water | 250 | 2–4 |
| -TE | — | — | | 155–175 | 8–12 |
| -TF | 525–545 | 4–12 | Hot water | 155–175 | 8–12 |
| LM26-TE | — | — | | 200–210 | 7–9 |
| LM28-TE | — | — | | 185 | ‖ |
| -TF | 495–505 | 4 | Air blast | 185 | 8 |
| LM29–TE | — | — | | 185 | ‖ |
| -TF | 495–505 | 4 | Air blast | 185 | 8 |
| LM30-TS | — | — | | 175–225 | 8 |
| **BS 'L'** | | | | | |
| 4L35 | 500–520 | 6 | Boiling water | 95–100 or room temperature | 2 5 days |
| 3L51 | — | — | | 150–175 | 8–24 |
| 3L52 | 520–540 | 4 | Water at 30–100°C | 150–180 or 195–205 | 8–24 2–5 |
| 4L53 | 425–435 | 8 | Oil at 160°C max‖ | — | — |
| 3L78 | 520–530 | 12 | Hot water | 160–170 | 8–10 |
| 2L91 | 525–545 | 12–16 | Hot water | 120–140 | 1–2 |
| 2L92 | 525–545 | 12–16 | Hot water | 120–170 | 12–14 |
| L99 | 535–545 | 12 | Hot water | 150–160 | 4 |
| L119 | 542 ± 5 | 5 | Boiling water or oil at 80°C | 215 ± 5 | 12–16 |
| L154 | 510 ± 5 | 16 | Water (50–70°C) | | 30 days |
| L155 | 510 ± 5 | 16 | Water (50–70°C) | 140 ± 10 | 16 |
| **DTD Specifications** | | | | | |
| 722B | | | | 165 ± 10 | 8–12 |
| 727B | 540 ± 5 | 4–12 | Water (80–100°C) or oil | 130 ± 10 | 1–2 |
| 735B | 540 ± 5 | 4–12 | Water (80–100°C) | 165 ± 10 | 8–12 |
| 5008B | — | — | | 180 ± 5 | 10 |
| 5018A | 430 ± 5 | 8 | Oil 160°C 1 h then oil at room temperature or air | — | — |
| or | 440 ± 5 | 8 | | | |
| then | 495 ± 5 | 8 | Boiling water | — | — |

†Single figures are minimum times at temperature for average castings and may have to be increased for particular castings.
‡'Hot water' means water at 70–80°C unless otherwise stated.
§The exact number of hours depends on the mechanical properties required.
¶The castings may be allowed to cool at 385–395°C in the furnace before quenching. The castings shall be allowed to stay in the oil for not more than 1 h and may then be quenched in water or cooled in air.
‖The duration of the treatment shall be such as will produce the specified Brinell hardness in the castings.

### 6.7.1.1

The stresses set up during solidification will vary depending upon the alloy; the size and shape of ingot; the speed of casting and the amount of cooling. If

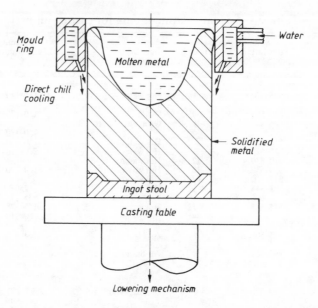

Fig. 65 — Schematic arrangement of direct chill (DC) semi-continuous casting process (*Courtesy Institute of Metals.*)

the latter are incorrect, the stresses set up can result in the splitting of ingots. This splitting may take place during casting or may occur during sawing to length. To avoid possible injury to personnel or damage to equipment, safe casting practices must be specified and closely adhered to. In addition, alloys which are very prone to splitting, such as the 7XXX series, should be stress relieved by heating to a temperature of about 450°C followed by slow cooling. This stress relieving treatment should be carried out as soon as possible after the completion of casting.

Methods for the reduction of internal stresses during casting include processes such as Pulsed Water Cooling [4], Aerated Water Cooling, or Atomised Spray Cooling, reduced casting speeds and increased mould skirt length, all of which can be employed to modify the rate of cooling and change the shape of the sump in the centre of the ingot.

### 6.7.1.2
While on the subject of safety precautions, attention is drawn to the hazards of molten metal and water explosions. Experimental work by Long and others [5] indicated that molten aluminium falling into water less than 2 inches (50 mm) deep did not produce violent explosions but metal was blown out of the tank in which the water was contained and scattered over the

surrounding area. Violent explosions occurred when the water depth was between 2 and 30 inches (50 and 750 mm). No explosions occurred when the depth of water was greater than 30 inches (750 mm) or when the metal fell through a distance of 10 ft (3 m) before entering the water. A number of safety measures have been based on these observations. The tests have also indicated that rusty surfaces increased the hazard of explosion, whereas surfaces cleaned free from rust and coated with selected grades of 'Rustoleum' paint reduced the hazard.

A series of four reports has been published by the sub-committee of the Health and Safety Executive (obtainable from HMSO) on continuous casting and high speed melting [6]. The first of these reports deals with 'Plant operation—aluminium/water hazards'; it is restricted to the casting of aluminium alloys in vertical machines using either semi-continous or continuous processes, and is confined to plant operation. The second report 'Causes and prevention of breakout during vertical semi-continuous casting of aluminium alloys' is concerned with practical measures which can be taken to reduce the risk of the breakout of molten metal during casting. The third report 'A warning and control system for continuous casting (as applied to copper alloys)' describes a device designed to provide a warning of a possibility of a run-out of molten metal before it actually occurs. The fourth report 'A study of the causes of molten metal and water explosions' contains a discussion of a research project at the University of Aston in Birmingham primarily directed to determination of the mechanism which initiates explosions when liquid metal and water mix in continuous casting plants. A number of tentative conclusions are reached, although they may need modification in the light of further experimental and theoretical work.

With the exception of magnesium, all the metals used in the pilot scale tests produced explosions. This work, therefore, as far as it had been completed, indicated that any molten metal which falls into water in suitable conditions will produce an explosion. Even as a tentative conclusion this has important implications for industrial plants.

No complete explanation of liquid–liquid explosions has yet been reached. Professor F. M. Page concludes 'The occurrence of explosions or disintegrations appears to be closely associated with the heat transfer reaching a critical rate under conditions of unstable or pulsation boiling'.

### 6.7.1.3  *Non-destructive testing*
In the development of casting practices it is usual to employ ultrasonic testing techniques to check for internal defects such as stress cracks and excessive porosity. Once satisfactory practices have been established a system of random checks may be instituted to establish whether the desired standard is being maintained.

For ingots which are to be used for aerospace components where small inclusions of non-metallics are unacceptable, a rigorous system of inspection is used.

### 6.7.2   Continuous casting
#### *6.7.2.1*
For the continuous casting of ingots by the *DC process* it has been found that horizontal casting units offer advantages over vertical units, one of these being that the operations and equipment are at ground level, with no necessity for deep pits. The process is used for the casting of small-diameter extrusion ingots but is not suitable for the casting of alloys liable to stress cracking on sawing to length.

#### *6.7.2.2*
The *Hunter–Douglas process* is used for the casting of a rectangular cross-section slab between two cylinders.

#### *6.7.2.3*
*Rigamonti* uses a wheel and endless belt to form the mould into which a continuous stream of metal may be poured.

#### *6.7.2.4*
*Hazelett* is the process in which metal is poured into the gap formed by two endless belts inclined at a slight angle to the horizontal. The product is flat and is cut to lengths for rolling since the casting speed is relatively slow (Fig. 66).

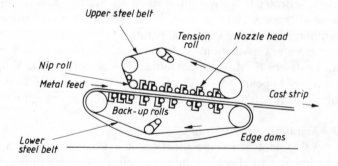

Fig. 66 — Diagram of the Hazelett machine (1958 model), (*Courtesy Institute of Metals.*)

### 6.7.2.5

*Rotary Strip Casting* (RSC) is the process where metal runs into a mould formed between the machined periphery of a forged or cast steel rotor and a mild steel belt, both being water cooled (Fig. 67).

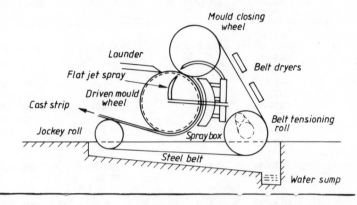

Fig. 67 — Schematic diagram of the Rotary Strip Casting Machine (RSC). (*Courtesy Institute of Metals*.)

These processes and developments of them to provide greater widths are confined to alloys with a narrow solidification range such as 1200, 3103, 3105, etc.

### 6.7.2.6

*Properzi* is a synchronised continuous casting and hot rolling process for the production of wire rod. The mould is formed between the machined periphery of a water cooled copper alloy wheel and an endless steel belt. Alloys cast by this process include 99.6–99.8 per cent Al (1060–1080), 1 or 2 per cent Mg, 0.5–0.7 per cent Si/0.6–0.8 per cent Mg (Almelec). Cracking is a difficulty with heavily alloyed wire compositions.

The as-cast structures resulting from these continuous casting processes are complex and are influenced by numerous factors inherent in the systems. Because of the differences between the structures of continuous cast stock and those of hot rolled DC ingot of similar composition, the mechanical properties of the two materials given the same cold rolling and annealing treatments will be significantly different.

## 6.8  HOMOGENISATION

Homogenisation (see Table 34) is a high temperature treatment to eliminate solute concentration gradients in the supercooled as-cast matrix and to take back into solid solution as many as possible of the crystals and intermetallics

**Table 34** — Typical working temperatures for melting, casting, homogenising and hot working aluminium and aluminium alloys

| Alloy design ation | Approx. melting range[a] (°C) | Casting temperature (°C) | Homogenising temperature (°C) | Initial hot working temperature (°C) |
|---|---|---|---|---|
| 1080A | 649–657 | | | |
| 1050A | 646–657 | 680–720 | 560–600 | 480–550 |
| 1200 | 643–657 | | | |
| 1350 | | | | |
| 2011 | 541–643 | | — | 400–440 |
| 2014A | 507–638 | | 480–490 | 400–440 |
| 2024 | 502–638 | | 480–490 | 400–440 |
| 2031 | — | 680–720 | 500 | 440–480 |
| 2117 | 554–649 | | 500–530 | 40–480 |
| 2618A | 560–649 | | 500–530 | 440–480 |
| 3103 | 643–654 | 680–720 | 580–620 | 480–520 |
| 3105 | 629–654 | | 530–550 | |
| 4043A | 547–632 | 650–700 | — | 460–500 |
| 4047A | 577–578 | | — | 460–500 |
| 5005 | 632–654 | | 530–550 | 440–480 |
| 5056A | 571–638 | | 380–420 | 400–420 |
| 5083 | 579–641 | | 380–420 | 400–420 |
| 5154A | 593–643 | 680–710 | 380–420 | 400–420 |
| 5251 | 607–649 | | 500–560 | 480–520 |
| 5454 | 602–646 | | 450–510 | 450–500 |
| 5554 | — | | 450–510 | 450–500 |
| 5556A | — | | 380–420 | 400–420 |
| 6061 | 582–652 | | | |
| 6063 | 616–654 | | | |
| 6063A | 616–654 | | | |
| 6082 | — | 680–710 | 560–600 | 430–500 |
| 6101A | 621–654 | | | |
| 6463 | 616–654 | | | |
| 7010 | — | | | |
| 7020 | — | 680–710 | 400–440 | 390–420 |
| 7075 | 477–635 | | | |
| Al–Li | | | | |

[a]Aluminium standards and data A74–75, Table 2.4.

formed or precipitated at the grain boundaries and their reprecipitation in a form which optimises the subsequent behaviour of the material during hot and cold rolling or during extrusion. It is a structure modification treatment [7].

Homogenisation causes structures to evolve rapidly twoards the state of thermodynamic equilibrium, without, however, eliminating all surviving effects of the casting process. Another effect of homogenisation is to relieve internal casting stresses.

Material which has been cast, homogenised and hot worked behaves differently from material from the same cast which has been homogenised, cooled to room temperature and then reheated for hot working. The effect of cooling back to 20°C is to precipitate the intermetallics and if the time of reheating prior to hot working is not long enough to take these back into solution the material will respond differently to hot working. The difference in temperature between the first and second treatment may lead to a fine precipitation of particles which control recrystallisation behaviour and the orientation of ears produced on deep drawing. Control of earing at an acceptable level depends on the temperature and time of homogenisation and on the percentage reductions in thickness between anneals.

## 6.9  ROLLING

### 6.9.1  Preparation for hot rolling

In recent years the trend has been to concentrate more and more of the hot rolling of aluminium at mills specialising in this type of commodity [8], the hot rolled stock then being distributed to other mills in the form of reroll stock for finishing by cold rolling.

DC cast ingot in thicknesses of up to 660 mm and widths of up to 2100 mm in lengths up to 5000 mm and weighing up to 19.5 t are being hot rolled to yield coils of 16–17 t and 2440 mm diameter. Expressed in terms of weight per unit width, coils may be as heavy as 8 kg/mm width.

The weights per unit width of coils produced on continuous casting and hot rolling units are governed by the capacity of the coiling and handling equipment, whereas coils rolled from ingot are governed by such factors as the dimensions of the preheat furnaces, the dimensions and power of the hot rolling mill, the lengths of the mill runout tables and the number of mills in the tandem hot finishing mill train.

Cladding of heat treatable alloys with either high purity aluminium for alloys of the 2XXX series, and aluminium–zinc for the 7XXX series, for the purpose of improving corrosion resistance, is performed during the hot rolling stage.

*1XXX Commercially pure aluminium*
DC cast ingots of this series have bands rich in iron and silicon at regular intervals across the face. This banding can give rise to coating problems on the surface of the hot mill rolls, resulting in defects which can be detected on the surface of processed sheet. Scalping (machining) to remove a thin

surface layer is normally carried out after the ingots have been homogenised but before reheating for hot rolling.

### 2XXX Aluminium–copper alloys

Alloys of this series have a generally thicker surface layer of constituent rich segregate which has to be removed by scalping. It is important to ensure that the scalped ingots have a fine machined finish since any surface defects will appear on the finished rolled product. Scalping is performed after homogenisation.

For the production of clad products, the machined surfaces are degreased by swabbing with a solvent. Plates to be used for cladding are scratch-brushed to remove the oxide coating resulting from hot rolling. The scratch-brushed plates are then placed against the ingot to form a sandwich. The plates are secured in position by steel bands and the composite is then preheated for hot rolling. The bonding of the cladding to the core does not take place until the first active pass of the composite through the hot mill, when a reduction of 123 to 25 mm may be given. The steel bands are removed before rolling commences. When dictated by strength considerations, one-side cladding may be used.

### 3XXX Aluminium-manganese alloys

The severe chill of the DC process retains a considerable amount of manganese in supersaturated solid solution. This can give rise to problems with coarse grain on annealing; hence the need for an homogenisation treatment to precipitate the manganese in a form which does not hinder the formation of recrystallization nuclei on annealing.

The segregate on the surface of 3103 alloy can be detected as a darkening of the surface of finished sheet. For non-critical applications this is considered to be acceptable but for applications such as container sheet which is to be decoratively printed, scalped ingots are required.

### 4XXX Aluminium–silicon alloys

Sheet in these alloys is not covered by British Standards. The American alloys AA 4343 (silicon 6.8–8.2 per cent) and AA 4045 (silicon 9.0–11.0 per cent) are used as cladding for brazing quality sheet (see Table 35). The core is either AA 3003 or AA 6951, both alloys having higher melting points than the cladding.

### 5XXX Aluminium–magnesium alloys

The addition of a small amount of beryllium (less than 0.008 per cent) reduces the amount of oxide which forms on the surface of the molten metal during casting, thus reducing 'cold shuts' and improving ingot surfaces. The presence of this element is also beneficial in reducing the amount of

**Table 35** — Nominal compositions of alloys used as cladding for brazing
quality sheet

| Designation | Percentage of alloying elements | | | |
| --- | --- | --- | --- | --- |
| | Silicon | Copper | Manganese | Magnesium |
| Cladding alloy | | | | |
| 4343 | 7.5 | | | |
| 4045 | 10.0 | | | |
| Core alloy | | | | |
| 3003 | | 0.12 | 1.2 | |
| 6951 | 0.30 | 0.25 | | 0.6 |

oxidation and discolouration of scalped surfaces during preheating for hot
rolling.

The rate of heating of the alloys with the higher magnesium contents
should be controlled so as to allow diffusion of low melting point constituents into the matrix. Homogenisation is normally carried out before
scalping.

### 6XXX Aluminium–magnesium–silicon alloys

Alloys of this series find use in applications where a combination of medium
strength, good ductility and formability, and weldability without serious loss
in properties is combined with good corrosion resistance. These alloys
require scalping before hot rolling but are used in the unclad condition for
most applications, though should cladding be required, a nominal cladding
thickness of 5 per cent per side of 7072 (Al–1 per cent Zn) is employed.

### 7XXX Aluminium–zinc alloys

The heat treatable alloys of this series require stress relief treatment before
sawing and scalping. Cladding with 7072 1 per cent zinc alloy is satisfactory
in affording improved corrosion resistance but where a combination of
improved corrosion resistance and a minimum loss in mechanical properties
is required, the heat treatable copper free Al–Zn-Mg-Cr-Mn alloy 7011 is
used (see Table 36).

### 6.9.2 Plate

Most plate commodities are finished rolled on the hot line. When a plate is
required in a temper other than soft or heat treated, the work hardened
temper is obtained by close control of the temperature of the slab and of the
amount of reduction in thickness during the final passes on the mill.

**Table 36** — Cladding alloys for heat treatable sheet and plate

| Designation | Forms | Cladding |
|---|---|---|
| Alcad 2014A | Sheet and plate | 99.3% Al *or* 0.7% Si, 1.2% Mg |
| Alclad 2024 | Sheet and plate | 99.3% Al |
| Alclad 3003 | Sheet and plate | Al–1% Zn |
| Alclad 5056 | Rod and wire | Al–1.2% Mg, 0.25% Cr, 2% Zn |
| Alclad 7075 | Sheet and plate | Al–1% Zn *or* Al–4.7% Zn, 1.2% Mg, 0.25% Mn, 0.12% Cr. |

Notes (1) Cladding may be specified for one or two sides.
(2) Cladding thickness — the nominal thickness will be greater than the average thickness determined by measurements at magnification of 100 on transverse microspecimens.
(3) The nominal thickness will be greater for thin sheet up to $1\frac{1}{2}$ mm thickness (5–10%) than for plate ($1\frac{1}{2}$–$2\frac{1}{2}$%).
(4) Alclad 5056 wire has nominal clad thickness of 20%.

Finishing operations include flattening, stretching and trimming to size, which for the heavier gauges will be by sawing and for the lighter gauges by shearing. Plates for heat treatment are edge trimmed to remove cracks which could act as stress raisers causing further cracking and splitting on quenching.

Solution heat treatment is performed in air heat treatment furnaces in a similar manner to that described later in the section on sheet heat treatment. Quenching is in water, and vigorous agitation of the quench bath is essential to prevent the formation of pockets of steam on the surface of the plate giving rise to soft spots due to a poor quench.

After quenching, plate is passed through a roller leveller or is given a few light passes on a rolling mill to remove distortion caused by quenching. It is then subjected to a controlled stretch of the order $2\pm\frac{1}{2}$ per cent to remove internal stresses and prevent serious distortion on subsequent machining [9]. Precipitation treatment is carried out after stretching and cutting to size.

When heat treated for critical applications, plates are subjected to ultrasonic inspection. Any defects present are more readily detected after solution heat treatment than in the as-rolled condition.

Representative specimens are selected from each heat treatment batch; they are selected in accordance with specification requirements and are subjected to mechanical tests.

### 6.9.3  Cold rolling

Reroll stock of the work hardening alloys, as supplied from hot rolling mills, is normally soft enough for cold rolling without need for preliminary anneal, whereas heat treatable alloys may require to be annealed.

The cold rolling as coil may be performed on four-high mills either on single stands or on multi-stand tandem mills. For heavy gauge material the hot mill slab may be cut into lengths for cold rolling on flat sheet mills. The reductions given during cold rolling are governed by factors such as gauge of starting stock and final thickness required, temper, metallurgical properties, grain size and earing characteristics on deep drawing. On four-high mills, (Fig. 68) reductions of up to 60 per cent in thickness per pass can be given to

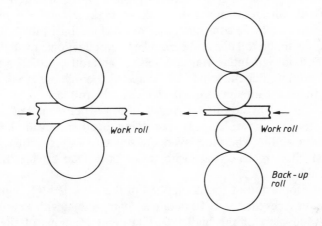

Fig. 68 — Diagrams of two-high and four-high rolling mills.

aluminium but the more highly alloyed materials which work harden more rapidly require smaller reductions per pass and annealing at more frequent intervals in the cold rolling schedule.

To produce intermediate tempers, at least one anneal is introduced into the rolling schedule for most products. For the purpose of grain size control it is necessary to observe both minimum and maximum reductions before annealing; the amounts vary depending upon the alloy.

### 6.9.4  Foil

Foil is defined as a cold rolled product of rectangular section and thickness not greater than 0.2 mm. The materials used for the production of foil are aluminium of various purities and alloys of the 3103 (1 per cent Mn) type. Although it is possible to roll foil in other alloys they do not find wide application.

The starter stock for foil rolling is generally about 0.5 mm thickness in the O or H4 temper. Rolling is performed on two-high mills or on four-high mills either as single stands or in tandem. Other mills such as the Sendzimir with small-diameter work rolls are also used for this purpose. Rolling reductions are usually limited to about 60 per cent per pass.

In sheet and strip rolling the reduction in thickness is effected by decreasing the gap between the work rolls. In foil rolling, and in particular when rolling the thinner gauges, the ends of the work rolls run in contact with one another. Under these conditions the effect of increasing the pressure on the roll necks is to increase the amount of roll flattening in the roll bite. To obtain a reduction in the thickness of the strand being rolled the factors which have to be controlled are: roll surface finish; for bright rolls reduce friction in the roll bite and permit heavier reductions; formulation of roll lubricant; unwind and rewind coil tension; and roll speed. The shape of the rolled product depends upon the shape of the ingoing stock and the control of thermal cambers across the width of the roll face.

The final pass for the thinner gauges of foil is performed as a double layer. To prevent the two layers from welding together they are held apart by a thin film of oil. When the layers are separated after rolling, the two surfaces which were in contact with the rolls will be bright; the inside surfaces will be matt.

A large proportion of foil is supplied in the annealed 'O' temper and batch-type furnaces are used. To obtain a clean product with a surface free from tacky deposits of the higher boiling point fractions of the rolling lubricant and free from indentations caused by bubbles of vaporised lubricant trapped between laps of the coil, the tension applied during coiling for anneal must be slack enough to permit the escape of vapour but not so slack as to allow inter-lap movement and surface abrasion.

## 6.10  ANNEALING AND HEAT TREATMENT

### 6.10.1  Annealing

Annealing to soften cold worked aluminium and aluminium alloys is normally undertaken at temperatures in the range 350 to 450°C but in commercially pure aluminium there is a pronounced fall in hardness when recrystallization occurs at temperatures around 250°C. In works' practice the annealing temperature is normally higher but the grain size of the annealed metal is greater the higher the annealing temperature above the level at which recrystallization occurs. The recrystallization temperature depends, of course, on the composition and the amount of prior cold working; thus aluminium of 99.996 purity is softened at temperatures as low as 100°C and superpurity metal may recrystallize at normal temperature. Fig. 69 shows the effect of annealing temperature on the hardness of

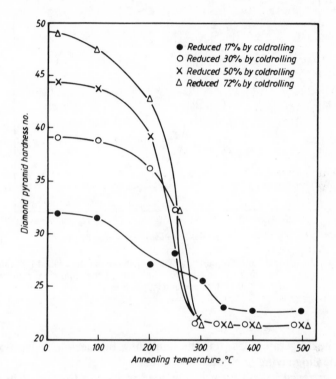

Fig. 69 — Effect of annealing temperature on diamond pyramid hardness of commercially pure aluminium with varying degrees of prior cold working.

commercially pure aluminium which has undergone various degrees of prior cold working. Conditions must be chosen to prevent the development of excessive grain growth which results in some loss of strength and the development of surface roughness (the 'orange peel' effect) when the metal is reworked. Correct fabrication and annealing conditions result in a uniform fine structure with a grain size around 0.02–0.10 mm diameter.

The effect of prior heat treatment may also be important, as it influences grain size and final hardness values, as illustrated in Fig. 70, owing to, for example, the quantity of impurities in solution before final rolling.

Commodities such as cut sheet or circles may be annealed in batch-type furnaces, the depth of the piles of the commodities being such as to ensure

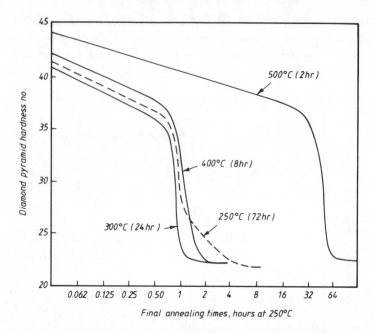

Fig. 70 — Isothermal annealing curves at 250°C of aluminium cold rolled 75%, after prior soaking at the times and temperature indicated. (The aluminium contained 0.14% silicon and 0.29% iron.)

uniform heating rates. Continuous annealing furnaces are also used for rapid annealing giving fine grain structures.

Most material in coil form is annealed in batch furnaces or in tunnel furnaces. Annealing a single strand is usually reserved for final annealing of thinner gauges of material.

When annealing coils in air, the rolling oil remaining on the surface and entrapped between the laps of the coil may not be completely vaporised and the higher boiling point fraction reacting with oxygen may leave brown breakdown products on the coil surface. High annealing temperatures are required to burn off these stains and in the case of magnesium-bearing alloys this results in black oxide stains. By annealing in a batch-type furnace having a controlled atmosphere with a low oxygen content and at lower temperatures than when annealing in air, the staining problems are reduced. After annealing in a controlled atmosphere the coils should be allowed to cool in the furnace to a temperature of less than 250 °C, otherwise residual oil on the coil is liable to breakdown on exposure to air.

Heat treatable alloys may be rolled in coil form but because of the need for close temperature control and for rapid quenching from solution heat treatment it is normal practice for all but light coils in narrow widths to be flattened after final rolling and to be cut into sheet lengths for solution heat treatment.

### 6.10.2  Heat treatment
#### 6.10.2.1  Solution heat treatment
At one time it was common practice to solution heat treat sheet in salt bath furnaces, but now it is more usual to use large forced air circulation furnaces for this purpose. The sheets are suspended from a frame and spaced to allow free circulation of the heated air. The rate of heating must be fast and the temperature uniformity during the soaking time should be within ±3 °C of the nominal value, since the optimum temperature is usually within only a few degrees centrigrade of that at which incipient fusion of low melting point constituents occurs (see Table 37). The time of soaking at temperature depends upon the thickness of the product being treated, ranging from a few minutes for thin sheet to a few hours for thick plate. For effective quenching the load should be lowered into the quench tank in a matter of seconds. The quenching operation leaves the sheet in a distorted condition owing to internal stresses, and after drying, the sheet is flattened through a roller levelling machine. It is then stretched by an amount of the order of 2 ± 1 per cent before being trimmed to width and length.

#### 6.10.2.2  Precipitation treatment
Precipitation, sometimes referred to as artificial ageing, is a low temperature thermal treatment, Table 37, designed to precipitate hardening phases from the solid solution resulting from solution treatment and quenching. Some alloys precipitate hardening phases at ambient temperatures, this process being very rapid during the first few days after quenching but may continue at a progressively slower rate over long periods. An alloy of this type is 2024, and other alloys such as 2014A or 6063 respond to natural ageing but when heated to temperatures of the order of 170–185°C for a period of a few hours give higher proof stress and ultimate tensile properties than those of the naturally aged material, though at the expense of lower elongation values and reduced formability.

Batch-type furnaces are used for precipitation treatment. The spacing of sheets for precipitation treatment is not as critical as for solution heat treatment, but the height of piles of sheet must be limited to ensure uniform heating throughout the furnace load.

### 6.11  EXTRUSION

Extrusion is a hot working process in which solid metal is forced by hydraulic pressure through an orifice in a die of alloy steel: it has been compared with squeezing toothpaste from a tube! To reduce die wear and obtain a longer die life, the orifice may be made of tungsten carbide or the steel may be

**Table 37** — Typical annealing, solution heat treatment and precipitation treatment temperatures for aluminium and aluminium alloys

| Alloy designation | Annealing temperature °C | Solution [12] heat treatment temperature °C | Temper | Precipitation treatment [13] Temperature °C | Time (h) |
|---|---|---|---|---|---|
| 1080A, 1050A 1200, 1350 | 360–400 | — | — | — | — |
| 2011 | | 525 ± 3 | T3 (TD) | Room | 48 |
| | | | T6 (TF) | 160 ± 3 | 12 |
| 2014A | | 505 ± 3 | T3 (TD) | Room | 48 |
| | 350–370† | | T6 (TF) | 175 ± 3 | 8 |
| 2024 | | 495 ± 3 | T4 (TB) | Room | 48 |
| 2031 | | 530 ± 3 | T6 (TF) | 175 ± 3 | 12 |
| 2117 | | 495 ± 3 | T4 (TB) | Room | 96 |
| 2618A | | 530 ± 3 | T6 (TF) | 180 ± 3 | 20 |
| 3103, 3105 | 400–425 | — | — | — | — |
| 4043A, 4047A 5005, 5056A | 350–370 | — | — | — | — |
| 5083, 5154A 5251, 5454 5554, 5556 | 350–370 | — | — | — | — |
| 6061 | | | T4 (TB) | Room | — |
| | | 530 ± 3 | T6 (TF) | 175 ± 3 | 8 |
| 6063 | | | T4 (TB) | Room | |
| | | 525 ± 3 | T5–T6 (TE–TF) | 175 ± 3 | 8 |
| 6063A | | | T4 (TB) | Room | |
| | | 525 ± 3 | T5–T6(TE–TF) | 175 ± 3 | 8 |
| 6082 | 350–370 | 535 ± 3 | T4 (TB) | Room | 120 |
| | | 535 ± 3 | T6 (TF) | 180 ± 3 | 10 |
| | | 530 ± 3 | T651 | 180 ± 3 | 7 |
| 6101A | | 525 ± 3 | T4 (TB) | Room | 120 |
| | | | T6 (TF) | Room | 175 ± 3 |
| 6463 | | 525 ± 3 | T4 (TB) | Room | 120 |
| | | | T6 (TF) | 170 ± 3 | 5–15 |
| 7010 | | 475 ± 3 | T7651 | 172 ± 3§ | 10 |
| 7020 | 290 ± 300‡ | 475 ± 3 | | | |
| 7075 | | 460 ± 3 | T73 | 110 ± 3 | 18 |
| | | | or | 120 ± 3 | 24 |
| | | | followed by | 177 ± 3 | 8 |
| Al-Li | | | | | |

†For complete annealing of material which has been heat treated, they should be heated at 400–425°C followed by cooling at about 15°C/h down to 300°C; below this temperature, cooling rate is not important.

‡For complete annealing, heat to 420°C, cool in air, reheat to 22 ± 3°C for 2 to 4 hrs.

§Heat to temperature at not more than 20°C/h.

squeezing toothpaste from a tube! To reduce die wear and obtain a longer die life, the orifice may be made of tungsten carbide or the steel may be coated with a refractory material such as silicon carbide. The principle is illustrated in Fig. 71 and as the cost of dies is comparatively low, sections designed for special purposes can be produced economically.

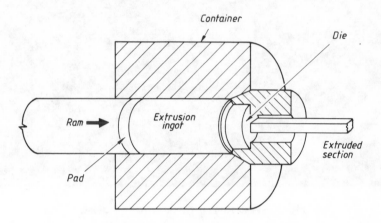

Fig. 71 — Principle of the extrusion process.

The range of alloys which can be extruded covers the soft high purity grades of aluminium through to the high strength 7XXX series. Some alloys which are difficult to frabricate as sheet products direct from as-cast ingot may be extruded to rectangular section and then rolled to thinner gauges but this is not common practice.

Extruded products may take the form of rod, bar, section and hollow section. Extrusion is also used to produce seamless tube stock for cold drawing to thin-walled seamless tube. Small-diameter wire rod may also be extruded for cold drawing to wire sizes. A selection of typical sections is shown in Fig. 72.

Thin-walled hollow sections are extruded from solid ingots through specially designed dies known as 'Bridge Dies'. The orifice defining the shape of the hollow section is formed by the die, machined to the external shape, and a short steel mandrel machined to give the internal shape. This mandrel is fastened to a shaped steel bar which bridges the orifice in the die. During extrusion the flow of metal is divided into two or more streams by the steel bridge; the metal streams (still solid) rejoin as they pass through the space between the die and the mandrel. The abutting surfaces are rejoined by pressure welding as they pass through the die. Strong, sound welds are formed with the softer alloys such as 1200 and 6063, particularly in sections

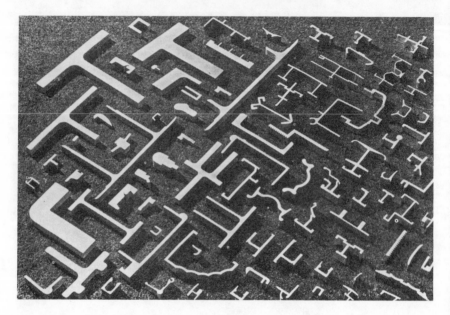

Fig. 72 — A selection of typical extruded sections.

of less than 3 mm thickness. This method of production is not recommended for stronger alloys of the 2014, 2024 or 7075 series.

### 6.11.1   Homogenisation for extrusion
When preheating for extrusion is carried out in long tunnel-type furnaces it is possible to combine homogenisation with heating for the extrusion process. However, the results are not likely to be as good as those obtained from ingot homogenised as a separate operation followed by cooling to about 20°C and fast reheating to extrusion temperature. Separate homogenisation treatment is required where induction heating or some other fast form of heating is employed for extrusion since the soak times are short and the temperatures too low to give efficient homogenisation. Temperatures for homogenisation and extrusion are given in Table 34.

### 6.11.2   Quenching at the press
The work hardening alloys of the 1XXX, 3XXX and 5XXX series leave the extrusion press in the soft condition but the temperature at which extrusions of the 6XXX series of alloys leave the extrusion die is within the solution heat treatment range. Sections of 6063 alloy in thicknesses of less than 3 mm, extruded at speeds of up to 50 m/min, may be air quenched at the press and will meet specification mechanical properties either as naturally aged (TB)

or, after being precipitation treated, (TE). Thicker sections require a more severe quench in water as they emerge from the press. Heavy sections extruded at relatively slow speeds may require separate solution heat treatment in a furnace, followed by quenching and natural or artificial ageing.

### 6.11.3 Solution heat treatment
Most alloys of the 2XXX series extrude at much lower speeds than those of the 6XXX series, and they also have narrow solution heat treatment temperature ranges; for these reasons these alloys are not suited for quenching at the press. Alloys of this series and those of the 7XXX series require furnace heat treatment. The furnaces may be vertical or horizontal with forced air circulation, and heating is followed by water quenching. As with plate and sheet the temperatures during heat treatment must be controlled within narrow limits. Finishing operations on extrusions include stretching, detwisting, correction of angles and dimensions (such as gap widths) and cutting to length.

### 6.11.4 Extrusion structures [10]
The structures of extruded products vary from end to end, the first metal through the die being virtually 'as cast'. For thin sections this will be a length of a few millimetres. As more metal is forced through the die, heavy deformation of the cast structure takes place. The as-cast grains are distorted and the constituents aligned in the extrusion direction, following the plastic flow lines as determined by the shape of the section and its symmetry about the extrusion axis. The grains are affected by the strain and the high temperature, and are successively hot worked and recrystallized (or recovered). For production purposes the extrusion temperature must be high enough to produce micro-recrystallization during extrusion but low enough to avoid melting any eutectics present. The extrusion speed will be as high as possible consistent with producing a good surface finish. This speed, for the higher strength alloys, is generally limited by the formation of transverse cracks on emergence from the die, owing to peripheral overheating. Sections having sharp corners are particularly sensitive to this form of failure.

To avoid excessive cooling of the ingot during extrusion, the press container liner is heated to a temperature about 20°C below that of the ingot. Too high a temperature will result in the oxide skin of the ingot flowing through the die to form a sub-surface layer on the section which on solution heat treatment or annealing, will be revealed as blistering or·flaking defects.

Macrographic examination of extrude stock reveals, a central zone of regular orientated texture, a fibrous distribution of insoluble intermetallics, surrounded by a peripheral zone (which may extend over much of the length of the section) of weakly defined (111) [100] texture, and an extreme outer zone (highly worked) showing no preferred orientation but containing finely distributed intermetallics. The very fine grain at the surface of the peripheral zone (immediately after extrusion) results, on subsequent solution heat treatment, in irregular giant grains.

Additions of Zr (0.1–0.2 per cent) in conjunction with Mn promote a reduction of the peripheral zone and retention of the extrusion texture during the homogenisation (solution heat treatment) preceding quenching [10].

The homogenisation of ingot preceding extrusion has the effect of taking into solution a large proportion of the particles that inhibit recrystallization. Thus, with all factors being the same, apart from homogenisation prior to extrusion, the material that has been homogenised must recrystallize more readily than that which has not received this treatment.

## 6.12  TUBE

Seamless tube is produced from ingot by extrusion over a central mandrel. The ingots may be cast as solid cylinders and then bored to take the mandrel or they may be pierced on the extrusion press by forcing a short cylindrical plug, slightly larger than the mandrel diameter, ahead of the mandrel. Should hollow cast ingots be used, the bore must be machined to remove oxide, segregate and casting imperfections before heating for extrusion. The annular space between the mandrel and the die defines the shape and dimensions of the extruded product which is known as 'tube bloom' when it is to be cold worked for the production of smaller diameters and reduced wall thicknesses.

Presses equipped to hold the mandrel in a fixed position in the die orifice (an 'arrested' mandrel) yield bloom of uniform wall thickness from end to end. Presses not so equipped use a tapered mandrel which advances through the die at the speed of the extrusion ram. Thus, the wall thickness changes progressively along the length of the bloom, being thickest at the front end. Longitudinal variations of this type can be corrected during subsequent cold drawing. Eccentricity of bore due to misalignment of the mandrel in the die will persist through to the finished drawn product. The actual amount will be determined by the eccentricity in the bloom and total drawing reduction.

For thin-walled tube and for tempers other than soft (O), the extruded bloom is cold drawn through a hardened alloy steel die or a tungsten carbide alloy die, over a bulb or plug, the dimensions of the drawn product being determined by the annular space thus formed. Reductions of cross-sectional area of up to about 30 per cent per pass may be given. For the harder work hardening alloys and the heat treatable alloys, annealing may be necessary after an overall reduction of about 60 per cent. Annealing may be conducted in horizontal batch-type furnaces.

The surfaces of the dies and bulbs must be highly polished and be kept well lubricated with copious supply of a good load-bearing lubricant. The drawn tubes should be degreased before annealing or solution heat treatment to prevent the risk of burning lubricant causing serious local overheating. Drawing through a die without the use of a bulb results in the inner surface of the tube becoming finely striated and rough, and only the outside diameter of the tube is reduced. The practice of drawing without a bulb (sinking) is sometimes used for the rectification of tube after solution heat treatment, a light pass of the order of 5 per cent reduction in diameter being employed. This practice leaves the tube in a stressed condition which is relieved by subjecting the tube to a stretching operation.

The solution heat treatment temperatures and times for tube are as for sheet products of the same alloy and thickness. To reduce the risk of inefficient quenching, tubes are preferably solution heat treated in vertical furnaces having the quench tank sited beneath the furnace.

Another method for the production of thin-walled tube from extruded bloom is by the use of a tube-reducing machine. The tube bloom is threaded over a long mandrel, fixed at one end. The free end is tapered. The reduction in wall thickness is effected by a pair of rolls, each having a tapered hemispherical groove machined around its periphery. The rolls and their housings, which are on slides, reciprocate backwards and forwards along the axis of the mandrel. The design of the grooved rolls is such that as they move along the tapered mandrel they form an annular gap, reducing in dimensions towards the tip of the mandrel. With each reciprocation of the rolls, the bloom is rotated and moved forward a slight amount and is compressed between the rolls and mandrel. Heavy reductions can be given by this method, which is more suited for the harder alloys than for pure aluminium and the softer alloys.

Seam welded tube is fabricated from the formed strip. This method of fabrication is normally confined to the work hardening series of alloys.


## 6.13  WIRE

Wire rod for drawing to wire sizes is usually 8–12 mm diameter according to the required diameter of the finished wire. The methods of production are by:

(a)  continuous casting, and rolling;
(b)  hot rolling;
(c)  extrusion.

The conductor grades of aluminium are readily cast by the continuous process but rods of the more complex alloys are produced either by hot rolling on a bar mill or by extrusion.

The mechanical properties and the grain structures obtained on annealing or solution heat treatment differ between hot rolled and extruded stock and for this reason the two types of stock require different processing schedules to yield similar properties.

Wire drawing is done on single wire blocks or continuous wire drawing machines; the wire drawing dies have an included angle of 7°, this being half as great as that normally used for copper [11]. If drawing is done with larger die angles, heating of the wire occurs which, as a result of the tempering (stress-relieving effect), is detrimental to its strength. This is of particular importance with aluminium for electrical conductors required in the hard drawn temper. Too great a die angle may give rise to internal fissures or 'cupping'.

## 6.14  FORGING

In general, forging of aluminium alloys requires considerable power applied relatively slowly. Forgings may be made from:

(i)   fine grained DC ingot;
(ii)  hot rolled stock from a bar mill or plate mill;
(iii) extruded stock.

The last-mentioned is commonly used, since it has a structure which has already been thoroughly worked to the centre. The larger diameters and sizes of extruded stock of the 2XXX and 7XXXX alloy type are usually machined, before forging, to remove the peripheral band of heavily worked structure which, on heating for forging and for solution heat treatment, would develop into coarse grain of unacceptable appearance and having low mechanical properties.

In slow compression the pressure required to deform aluminium and its alloys increases with increase in alloying content, those of the 6XXX series of alloy being only slightly higher than those for 1XXX alloys whilst increasing copper contents of the 2XXX type give rise to much higher power consumption. The strongest influence on power requirements is exerted by increase in magnesium content of the 5XXX alloy series. Observations show that power consumption increases very considerably with increasing speed of compression.

Behaviour under hammer forging is similar to that in pressing. In die forging, good lubrication with cylinder oil of high flash point considerably eases deformation.

The crank press, the hydraulic press and friction screw presses with constant speed during deformation of the stock show the lowest and thus most suitable forging speed for aluminium. If drop-hammers are used, the drop should be be reduced in favour of a correspondingly greater weight.

The properties of forgings depend upon the type of forging stock, the homogenisation treatment, the type of preworking (upsetting and reducing parallel to the original axis of the stock or at right angles to the stock or triaxial working). Elongation values are improved if material is cogged (upset) along the three axes (longitudinal, long transverse and short transverse) before forging. The fatigue limit is also improved for 2014A alloy given this type of treatment.

Heating for forging is performed in furnaces equipped to give uniform heating, the forging ranges being limited to some 40–50°C. The solution heat treatment temperatures are the same as those for extrusions of the same alloy. Details of these treatments are given in Tables 34 and 37.

## REFERENCES

[1] Van Lancker, M., *Metallurgy of Aluminium Alloys*, 1967, Chapman & Hall, London, p. 191.

[2] Aluminium — the Production of Castings — Wall Chart No. 4, Teachers Notes, Aluminium Development Association.

[3] *The Properties of Aluminium and its Alloys*, p. 65 — Table 25, Aluminium Federation.

[4] Pulsed Cooling, British Patent 1,045,423.

[5] Long, G., Explosions of molten aluminium in water — causes and prevention, *Metal Progress*, May 1957.

[6] *First Report*: Operational safety during vertical semi-continuous and continuous casting of aluminium.
*Second Report*: Causes and prevention of breakout during vertical semi-continuous and continuous casting of aluminium alloys.
*Third Report*: A warning and control system for continuous casting (as applied to copper alloys).
*Fourth Report*: A study of the causes of molten metal and water explosions.
Continuous Casting and High Speed Melting, sub-committee of Joint-standing Committee on Health, Safety and Welfare in Foundries, HMSO.

[7] Carmel, L. J., King, F. and Salt, G., Hot rolling of sheet and strip: aluminium and aluminium alloys, *Metals Technology*, August 1975, p. 313.

[8] Van Lancker, M., *Metallurgy of Aluminium Alloys*, 1967, Chapman & Hall, London, pp. 207, 211, 212.

[9] Forest, G., Internal or Residual Stresses in Wrought Aluminium Alloys and their Structural Significance, **58**, No. 520, April 1954, p. 261.

[10] Van Lancker, M., *Metallurgy of Aluminium Alloys*, 1967, Chapman & Hall, London, p. 231.

[11] Von Zeerleder, A., *The Technology of Aluminium and its Alloys*, pp. 251–261, English Translation, 1948, High Duty Alloys Ltd.

[12] *Aluminium Standards and Data* 1974–75, Table 2.4, The Aluminium Association.

[13] *Heat Treatment Data for Casting Alloys*, Table 26, Aluminium Federation.

**FURTHER READING**

Dassell, J. L., Inclusion removal from molten aluminium alloys with rigid media filter, *Light Metals AIME*, 1973, pp. 459–467.

Blayden, L. C. and Brondyke, K. J., In-line Treatment of Molten Metal. *AIME*, 1973, pp. 493–503.

Linde Spinning Nozzle Inert Filtration Process, US Patents 3,227,547, 3,870,511 and 3,980,742.

Method of Filtering Molten Light Metals, US Patent 3,537,987.

Ceramic Foam Filters, US Patent 3,947,363.

Improvements in or Relating to the Treatment of Liquid Metal, British Patent Specification 1,316,578.

Removal of non-metallic Constituents from Liquid Metal, British Patent Specification 1,367,069.

Apparatus and procedure for Treatment of Molten Aluminium, Canada Patent Office 554,853.

Introducing a Grain Refining or Alloying Agent into Molten Metals and Alloys, US Patent 3,634,075.

Servier, P. E., Dore, J. E. and Petersen, W. S., Considerations in the evaluation of aluminium master alloys, *J. Metals*, Nov. 1970, pp. 37–46.

Doherty, R. D. and Martin, J. W., The effect of a second phase on the recrystallization of aluminium-copper alloys, *J. Inst. Metals*, **91**, p. 332.

The metallurgy of light alloys, Conference Proceedings 1983, 318 pp.

Aluminium–lithium alloys, Conference Proceedings 1986, 640 pp.

Aluminium Technology '86, Conference Proceedings 1986, 850 pp.

Published by the Institute of Metals, London.

# 7

# Manufacturing processes

Aluminium can be fabricated in much the same way as other metals and in general the same tools can be used, although good quality tooling with well polished surfaces is desirable. A certain amount of care is also necessary in handling aluminium when surface finish is important.

As aluminium is a relatively soft material, any dirt, corrosion products, swarf, filings of other metals, etc., can easily cause surface damage, especially during transit; such particles may be held in any oil or grease film and may even become embedded in the metal. Sheets can be ruined by being dragged over one another by careless operators; they should always be lifted clear, as should long tubes and extrusions.

## 7.1 BENDING [1]

The behaviour of metal under bending, as demonstrated by the routine laboratory bend test, is often taken as a very broad guide to general workability. Fig. 73 illustrates this test.

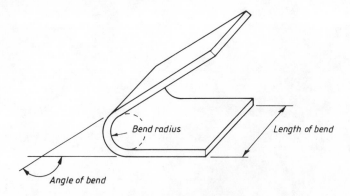

Fig. 73 — Terms used in bending. The radius of bend may be defined as a multiple of the thickness ($t$) of the specimen or of the workpiece.

Typical values obtained in different thicknesses of sheet in various alloys are given in Table 38. Other things being equal, the smaller the bend radius,

**Table 38** — Typical bend radii* for aluminium alloy sheet of various thicknesses

| Alloy designation | Temper | 180° bed test radii Sheet thickness — nm | | |
| --- | --- | --- | --- | --- |
| | | 3.2 | 1.6 | 0.9 |
| 1200 | 0 | Flat | Flat | Flat |
| | H2 | Flat | Flat | Flat |
| | H4 | Flat | Flat | Flat |
| | H6 | $\frac{1}{2}$t | Flat–$\frac{1}{2}$t | Flat–$\frac{1}{2}$t |
| | H8 | $\frac{1}{2}$–1t | $\frac{1}{2}$t | $\frac{1}{2}$t |
| 3102 | O | Flat | Flat | Flat |
| | H2 | Flat | Flat | Flat |
| | H4 | Flat–1t | Flat | Flat |
| | H6 | $\frac{1}{2}$–1t | $\frac{1}{2}$–1t | Flat–1t |
| | H8 | 1–2t | 1–1$\frac{1}{2}$t | 1–1$\frac{1}{2}$t |
| 5251 | 0 | Flat–1t | Flat | Flat |
| | H4 | 1t | $\frac{1}{2}$–1t | $\frac{1}{2}$t |
| 5154a | 0 | 1t | Flat | Flat |
| | H2 | 1$\frac{1}{2}$–3t | 1t | $\frac{1}{2}$–1t |
| 5083 | 0 | Flat–1t | Flat | Flat |
| | H2 | 1–2t | 1t | $\frac{1}{2}$–1t |
| 6082 | 0 | Flat | Flat | Flat |
| | TB | 1–2t | 1–1$\frac{1}{2}$t | $\frac{1}{2}$–1t |
| | TF | 2–3t | 1$\frac{1}{2}$–2t | 1–1$\frac{1}{2}$t |
| 2014a | 0 | Flat–1t | Flat | Flat |
| | TB | 2$\frac{1}{2}$t | 2t | 1$\frac{1}{2}$t |
| | TF | 4t | 3t | 2t |
| 2024 | 0 | $\frac{1}{2}$t | Flat | Flat |
| | T4 | 4t | 3t | 2$\frac{1}{2}$t |
| 7075 | 0 | $\frac{1}{2}$t | $\frac{1}{2}$t | Flat |
| | T6 | 4–5t | 3–4t | 2–3t |

*Values of bend radii given in terms of sheet thickness 't'.

the bigger the bend angle, and the larger the bend length, the greater is the risk of fracture.

Heat treated material work hardens more rapidly than the work hardening series of alloys, and for equal elongation values the heat treated material cannot be bent to as small a radius; for example, 1.6 mm thick 2014 alloy sheet in the soft ('O') condition can be bent flat but in the TB temper the minimum bend radius is twice the thickness of the sheet.

'Springback', which occurs with most metals when the bending pressure is released, represents the elastic deformation that occurs in addition to the plastic deformation that gives sheet its permanent set. It is greatest with material having a high elastic limit and increases with decreasing metal thickness. Methods of compensating for this effect include overbending, the amount being determined by trial and error.

For straight-line bending thin sheet over small radii the simplest tool is the hand folder; thicker sheet can be bent in the press brake. It is an advantage if the brake can be operated at different speeds so that the more critical operations and those demanding greater accuracy can be performed more slowly than the general run of bending work.

The bending of aluminium in the press brake requires a substantially different technique from that for steel. Unless the material is controlled throughout the bending operation there is a risk of forming peak radius in advance of the punch as it descends. The formation of peak radius can be avoided by having a spring loaded pressure pad or other pads used in conjunction with spring loaded pressure pads.

Various rotary flanging and edging machines used for turning up straight, irregular or curved edges on sheet steel are equally suitable for aluminium. Parts of wide curvature such as the walls of cylindrical vessels are most conveniently produced by roll formers, three roll machines being commonly used.

Continuous roll forming may be used for making long sections from sheet or coil of required developed width. Several pairs of rolls in tandem progessively bend the metal to the desired shape. There is practically no limit to the shapes which can be produced in this way provided the rolls at each station are carefully designed to bring the sheet gradually to its final shape. Seamed or butt-jointed tubes can be produced in addition to the more common angle or channel section in corrugated sheet. Rolls must be as highly polished as possible.

## 7.2  SPINNING [2] (Fig. 74)

Because of its excellent ductility and its freedom from excessive work hardening, commercially pure aluminium can be spun more easily than any other sheet metal. When a spun component of high strength is required, the wide range of aluminium alloys offers a choice of materials combining some of the pure metals ease of spinning with the strength required. Very broadly, the bigger and more intricate the article to be produced, the more economical it will be to spin it rather than to produce it by pressing and drawing. Whereas 1200 alloy temper 'O' can be worked into complex shapes without intermediate annealing, 3103 1.25 per cent Mn temper 'O' and 5152 (2 per cent Mg–0.25 Mn) alloy temper 'O' may require intermediate annealing in the production of very deep or intricate shapes. For the heat treatable alloys

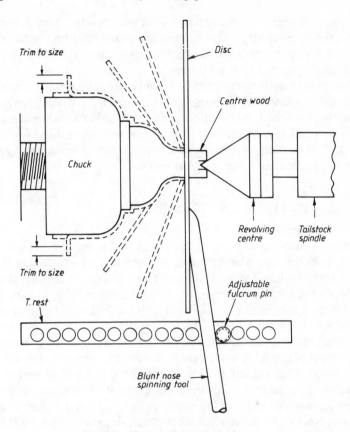

Fig. 74 — The principles of spinning.

which work harden rapidly, annealing is more essential. As an alternative to intermediate annealing it might be preferable to solution heat treat the blank after it has been roughly spun, and immediately finish spinning before the metal has naturally age-hardened to any appreciable extent. In some conditions, where the material is thick, for example, it might be advisable to spin the metal hot using gas flames to heat the metal to about 200–300°C.

The portion of the blank representing the base (or the crown) is usually untouched by the spinning tool and it will not become work hardened. If necessary it can be hardened later by some form of working such as hand hammering. When heat treatable alloy is used, the whole article is hardened by solution heat treatment and quenching.

Because of its high ductility, aluminium can be spun at faster speeds than most other metals: a peripheral speed of about 900 m/min is recommended. Although this will depend on the alloy, temper, the shape of the finished article and, not least, on the experience of the operator, 1500 m/min should in no case be exceeded. The head stock of the spinning lathe is usually designed for several speeds to accommodate these variables, the spindle speed ranging from 200 to 2000 rpm.

The chucks or formers used in spinning aluminium are commonly of hardwood, seasoned beech being often preferred, but when large quantities of spinnings are required, mild steel, cast iron and cast aluminium are often specified. Chill cast aluminium formers are very often used for spinnings over 0.75 m in diameter, being as completely rigid yet much lighter and easier to handle than other metal formers.

## 7.3  DEEP DRAWING AND PRESSING [3]

### 7.3.1  Introduction

The aim in drawing is to form the blank to the required shape without causing it to thin, wrinkle or fracture. The pressure exerted by the blank holder should, therefore, be just sufficient to enable the metal to move parallel to the surface before it enters the die impression; greater pressure would stretch the metal and possibly fracture it; less would relax control of it to the extent of permitting wrinkles to form in a radial direction. The portion of the top flange of the shell, that forming the base, is virtually undisturbed, but that part of the blank undergoing the most movement is that forming the sides. In a circular shell the movement of the metal is uniform in all diameters; in a reactangular shell the greatest movement occurs at the corners from which some of the metal tends to flow to the sides where, like the base, little or no movement takes place—it may, in fact, be little more than simple bending.

The aluminium alloys most commonly deep drawn and pressed are those of the work hardening groups. Commercially pure aluminium and 1.25 per cent Mg (3103) alloy are the most easily drawn and most widely used, especially in the production of holloware. The other widely used alloy of the work hardening type, 5152 (2 per cent Mg–0.25 percent Mn), is often specified where increased strength and rigidity are needed and its good resistance to sea water corrosion is an advantage.

The maximum reduction in diameter that should be attempted in a single draw to avoid fracturing the metal has been given as 48 per cent for the work hardening alloys and 40 per cent for those of the heat treatable type in the freshly solution heat treated condition. It has been laid down that as a general rule, more than one operation is necessary to draw to size a shell in which the depth is more than about 70 per cent of the diameter; this limitation applies to free drawing in which there is little or no ironing (that is, thinning of the metal in the walls).

For aluminium and its alloys, the radius over which the metal moves into the die should not be less than about four times, or more than fifteen times, the thickness of the blank: the same minimum ratio applies to the punch radius (Fig. 75).

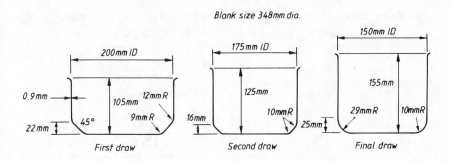

Fig. 75 — Deep drawing—sequence of operations for cylindrical shapes.

So that the metal will move as freely as possible and not tear or thin, the clearance between the punch and die must be appreciably greater than the thickness of the blank, yet not too great that control is reduced. The proper clearance should range from 107 to 120 per cent of the thickness and should increase with each successive draw. Table 39 gives the average values of

**Table 39** — Average reduction for first draw, for material conforming to BS General Engineering Specification 1470, Alloys 1200 and 3103, 'O' temper

| | Thickness/diameter ratio | | |
| | Single-action press | | |
| Double-action press | With blank holder | Without blank holder | Approx. reduction in diameter (%) |
| --- | --- | --- | --- |
| 0.0015 | 0.0025 | 0.02 | 35 |
| 0.0025 | 0.0030 | 0.03 | 40 |
| 0.004 | 0.0045 | 0.04 | 45 |
| 0.005 | 0.0055 | 0.05 | 48 |

reductions for first draws for 1200 and 3103 alloys in the soft annealed 'O' temper and Table 40. the scale of reductions for redrawing.

**Table 40** — Scale of reduction for redrawing. For material conforming to BS 1470, alloys 1200 and 3103, 'O' temper

| 1st draw from blank (%) | Redraws | | | |
| | 2nd draw (%) | 3rd draw (%) | 4th draw (%) | Total reduction (%) |
| --- | --- | --- | --- | --- |
| 35 | 20 | 15 | 10 | 60 |
| 40 | 23 | 18 | 13 | 67½ |
| 45 | 27 | 22 | 16 | 73½ |
| 48 | 30 | 25 | 18 | 77½ |

When the point is reached at which the metal is too hard for further cold working, it is annealed. Workability is seldom fully restored, particularly after several draws. The reason is that all parts of the metal will not have been worked to the same degree; therefore recrystallization that accompanies annealing will produce a variable grain size: areas having received the least work will have the largest grain and will tend to be the weakest under stress. For this reason the sheet should be worked as much as possible before drawing (e.g. H2 or H4 temper).

Deep rectangular and irregular shaped shells are more difficult to draw than cylindrical ones. With alloys of inferior ductility, the drawing of deep boxes with sharp corner radii becomes less easy than for 1200 or 3103 type alloys and is impossible with high strength alloys.

Factors other than thicker corner radii which control the permissible reduction are the size and shape of the blank, and the design of the tools. The corners of the shell are the areas where most drawing occurs and care must be taken to ensure that the blank provides sufficient metal at these points but not so much that it cannot be controlled.

### 7.3.2 Drawing speed
The shape of the part and the ductility of the alloy greatly influence the speed at which drawing can be done. Generally the risk of wrinkling or cracking is

intensified as the speed is increased. A comparison with the speeds of drawing other metals is given in Table 41.

**Table 41** — Open drawing speeds for different metals

| Material | Single-action presses m/min | Double-action presses m/min | Ironing m/min |
|---|---|---|---|
| Stainless steels | 12 | 9 | 6 |
| Low carbon steels | 15 | 12–15 | 7.5 |
| Zinc | 45 | 12–15 | — |
| Aluminium | 48 | 15–24 | 18 |
| Copper | 45 | 18–25.5 | 15 |
| Brass | 60 | 24–30 | 21 |

The most obvious result of employing an excessively high drawing speed is to push the bottom out of the shell; another is scoring, due to the generation of excessive heat which weakens or destroys the lubricant and causes undue friction; another is rapid wear of the tools and scratching of every blank brought into contact with them.

### 7.3.3  Lubrication
Mineral oils and compounded mineral oils are very widely used: they have the advantage over water soluble oils in that a single application will provide sufficient lubrication for several redraws.

### 7.3.4  Ironing
While the aim of deep drawing is to achieve the required reduction in diameter from the blank to finished shell without reducing the thickness of the metal, articles that are thicker at the base than the walls can be produced on the draw press by deliberately thinning the metal as required by simple ironing in the press.

Ironing necessarily strain hardens the metal to a greater extent than free drawing, yet reductions in wall thickness of up to 40 per cent can be accomplished in a single operation, particularly in the low strength alloys. It is important to recognise that wall thickness cannot be severely reduced at the same time as diameter; therefore, in common practice, ironing operations follow deep drawing.

The principle of ironing is similar to tube drawing in that the clearance between the punch and the die is less than the metal thickness before the operation. In ironing, the clearance between the punch and the die should be 90 per cent of the final metal thickness but tube drawing speeds should not be attempted. Speeds lower than in deep drawing are recommended, particularly on mechanical presses; hydraulic presses offer greater control over the operation and are to be preferred. The percentage reduction in wall thickness should be decreased with each draw; in low strength alloy the reduction would be about 40 per cent in the first operation and about 20 per cent in the second.

## 7.4   DROP-HAMMER FORMING [4]

The drop-hammer is used in the aluminium industry almost exclusively for hot forging, but in other industries, especially aircraft production, it provides a speedy, economical and often indispensable means of forming aluminium sheet into a variety of shapes.

The chief advantage of drop-hammer forming that makes limited production runs economically possible is that soft metal dies can be used. Bottom dies are normally cast in a 96 per cent Zn–4 per cent Cu alloy and punches in the same alloy or in lead alloyed with 10–12 per cent antimony. As there is a considerable difference between the melting points of lead and zinc and also the degree to which these metals contract, it is possible to cast the punch from the bottom die, which is very convenient. The contraction of the punch on cooling usually provides sufficient space between the tools to allow for the thickness of the sheet being formed but care should be taken to ensure correct mating of the tools and accurate mounting in the hammer. Both punch and dies are polished to give a good working face.

Simple components can often be produced at a single blow whereas others require several of varying intensity. In drop-hammer forming certain shapes, wrinkles are very difficult or impossible to avoid. Wrinkles may be avoided, or at least reduced, by adopting one of several methods of controlling the metal. Pressure pads, or draw-rings, are widely used to determine the stages in which the metal is to be formed. They are distance pieces of plywood or rubber about 18 mm thick, routed roughly to the appropriate shape; these are piled over the blank on the die face and one or more being removed after each draw, thus allowing the punch to move the metal into the die a little at a time.

## 7.5   RUBBER DIE PRESSING [5]

The use of rubber in the forming of aluminium sheet is given its fullest scope in rubber die pressing by what is normally known as the Guerin process. In this method, only a male die is required, the function of the female die being performed by a thick rubber blanket housed in a strong steel container. For simple forming, flanging and drawing operations with which can be incorporated shearing, blanking and piercing, a rubber die press is an indispensable item of equipment in many production departments working with aluminium sheet.

The presses used for rubber die forming are normally worked at 11.6 N/ $mm^2$ (0.75 tons/in$^2$), and are suitable for heavy gauges; presses working on the same principle with a slightly modified die design exert about half as much pressure, and are particularly useful in forming sheet between 2 and 0.9 mm thickness. (Fig. 76).

By selection of rubber of the correct hardness it should be possible to form some 20,000 parts of average size and shape without appreciable wear of the rubber pad.

Dies used in the rubber die press are usually simple form blocks made of steel, cast iron, hardwood, impregnated hardwood, fibre or zinc (96 per cent zinc–4 per cent copper, cast at 450°C, requiring a contraction allowance of 1 in 120). Work given to a rubber die press seldom calls for dies deeper than about 38 mm.

In all dies the forming radii should be as large as possible to prevent cracking of the sheet and undue wear on the rubber; a minimum radius of three times the thickness of the sheet is often specified. Generally 'backing off' the die face to provide 4° over bending adequately compensates for the elastic recovery 'springback' of the sheet.

For lubrication, the oils and greases normally used in the power press are unsuitable, since they damage the rubber. French chalk or fine graphite, dusted on the sheet before pressure is applied, is very successful.

The strong heat treatable alloys in the fully soft 'O' condition are among those most commonly formed by this method, although advantage is taken of the temporary softness of the 'as quenched' whenever possible, thus avoiding possible distortion of the formed part on quenching.

The work hardening alloys, also, are commonly rubber die formed when fully soft, but it is sometimes an advantage to form them in the H2 or H4 temper.

## 7.6  STRETCH FORMING [6]

The stretcher press forms entirely by tension, elongating the metal beyond its elastic limit. Springback is practically non-existent: the contour of the die face is made to correspond exactly to that of the finished shape. During stretching, the sheet decreases in thickness from about 3 to 5 per cent, this proportion being roughly 40 per cent of the total elongation value. To ensure effective distribution of strain, the blank is generally rectangular, whatever the shape of the part, and after it has been stretched, the boundary of the

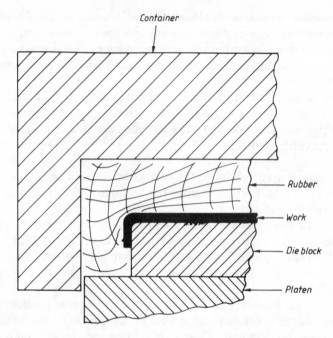

Fig. 76(a) — Rubber die pressing—action of rubber pad under compression.

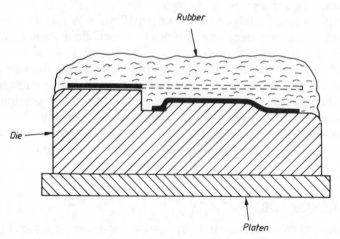

Fig. 76(b) — Rubber die pressing — combined blanking and forming tools.

component is marked out using a template. A disadvantage of stretch forming is the fairly high scrap loss, for the area of sheet grip in the jaws of the press has to be cut off as waste.

The main requirement in alloys to be stretch formed is the uniform stretching of the material up to the point at which general elongation is exhausted and necking begins; the most suitable material is, therefore, that having a high elongation value and a fairly wide plastic range (indicated by the difference between the proof stress and ultimate stress). Further, the elongation value will be a more reliable value of formability if measured on longer gauge lengths than normal when the local elongation (representing necking) is smaller in relation to the total elongation figure.

As in other methods of forming, more than one operation may be necessary, possibly with intermediate annealing to restore in some measure the original ductility of the metal. When intermediate annealing is necessary, excessive grain growth may be encountered, revealing itself in an unsightly 'orange peel' effect. By a practice such as a sub-critical temperature anneal at a temperature lower than that needed for recrystallization, or the germination of new grains, the orange peel effect may be eliminated.

Lubrication of the die face is usually necessary, the choice of lubricant being governed by the die material. Water-bearing lubricants should not be used with wooden dies because of the risk of warping. Excessive lubrication, especially in forming shallow parts, is to be avoided since there is a risk of trapping lubricant beneath the work piece and creating bulges which are difficult to remove.

## 7.7  HAND FORMING [7]

The art of hand working aluminium sheet is widely practised for small production runs or prototypes which do not justify the sinking of dies. The equipment that has been developed and used through the years for hand working steel, copper, silver and other metals is equally suitable for aluminium, and the technique used is broadly the same. Severe forming of any of the aluminium alloys is usually begun with the metal in the fully soft condition. The hammers used with aluminium are generally lighter than those used with harder metals.

## 7.8  BLANKING AND PIERCING [8]

The most common method used to cut blanks from aluminium sheet is that of a punch and die, the main difference from tools used in blanking other metals being the clearance required between punch and die which is governed by shear strength and metal thickness. For aluminium, the clearance is roughly one-tenth of the thickness of the sheet (about double that for steel).

In the press blanking operation, true cutting takes place for only about one-third of the sheet thickness, being followed by fracture which begins at

opposite sides of the sheet as the metal is stressed to breaking point. The
importance of having the proper clearance between top and bottom tools is
that if it is too small, the fractures fail to meet, producing a ragged edge, and
excessive strain is put on the punch and die; on the other hand, if the
clearance is unduly large, the metal tends to be drawn rather than cut and
again the result is a rough edge and excessive strain on the tools. Punch and
die clearances for blanking aluminium alloys are given in Table 42.

**Table 42** — Punch and die clearances for blanking aluminium alloys

| Sheet alloy | | Clearance | |
| --- | --- | --- | --- |
| Designation | Temper | On a side | On the diameter |
| 1200 | O | $0.10t$ | $0.20t$ |
| | H4 | $0.12t$ | $0.24t$ |
| | H8 | $0.14t$ | $0.28t$ |
| 3103 | O | $0.10t$ | $0.20t$ |
| | H4 | $0.12t$ | $0.24t$ |
| | H8 | $0.14t$ | $0.28t$ |
| 2014A | O | $0.13t$ | $0.26t$ |
| | TF | $0.16t$ | $0.32t$ |
| 6082 | O | $0.11t$ | $0.22t$ |
| | TB | $0.12t$ | $0.24t$ |
| | TF | $0.14t$ | $0.28t$ |
| 7075 | O† | $0.13t$ | $0.26t$ |
| | TF† | $0.16t$ | $0.32t$ |

$t$ = thickness of sheet.
†Thick sheet (greater than $2\frac{1}{2}$ mm) and plate should not be blanked in the TB or TF temper
because of the risk of cracked edges. This material should be sawn or machined to shape.

The hardened die determines the size of the blank, and the soft steel
punch the size of the hole; in blanking, therefore, the clearance value is
subtracted from the punch diamter and, in piercing, it is added to the
diameter of the die. To allow the blank to drop freely, the die walls are
tapered away from the cutting edge at an angle of about $\frac{3}{4}°$.

Blanks that are too big to be cut in a press, or are required in quantities
too small to justify the cost of press tools, are often produced on a circle
cutter. The blanks are cut on this machine by revolving flat sheet horizon-
tally through 360° between cutting wheels. The minimum thickness of sheet
that can be handled in this way is about 0.45 mm. For thick material and

complex shapes the blanking may be carried out by routing. In this operation the sheet is cut by a vertical high speed milling tool mounted on the end of a swivelling arm and guided by a template.

The main differences between blanking and piercing (or 'perforating' when several holes close together are punched simultaneously) are that the die face is ground flat to obviate distortion of the strip, while the face of the punch is ground to an angle to reduce the shearing load on the press. The clearance should be no more than 5 per cent of the metal thickness and should be added to the punch diameter which represents the diameter of the hole. As in blanking, the holes in the die should be tapered to allow the slugs to drop out.

When large numbers of holes are to be punched in the same sheet it is quicker to use a gang punch which will do the piercing in a single operation. Such tools can be made by adapting a press brake or shear by fitting standard punches. Less strain will be put on the press and the holes will be aligned more accurately if the punches are of slightly different lengths.

It is quite common for blanking or piercing to be carried out at the same time as drawing, and it is not uncommon for all three operations to be performed together.

## 7.9 SUPPLEMENTARY OPERATIONS [9]

It very often happens that a hollow vessel formed on the draw press requires one or more operations to give it its final shape.

### 7.9.1

Expanding or bulging can be carried out using a variety of media—segmented tubes, rubber, oil or water, depending to some extent on the shape, thickness and hardness of the metal.

Expanding by means of rubber pads is efficient and widely used. The rubber pad mounted in either the punch or the die is of a hardness suited to the severity of the operation. Experiment has shown that the grade most satisfactory for severe forming operations has a strength of about 0.69 $N/mm^2$ (10 $lb/in^2$) for each 1 per cent of compression. A 20 per cent compression should, however, be regarded as a maximum.

Water and oil are not as widely used as rubber as expanding media. A common method of expanding aluminium shells is by spinning, particularly if small quantities only are required.

### 7.9.2

Spinning also provides a good method of reducing the diameter of aluminium shells (i.e. contracting or necking—the reverse of expanding). In this operation the metal is stressed entirely in compression and is thickened as the tool moves over it. It is recommended that the reduction in diameter for each operation should not exceed 8 per cent for metal in the harder tempers or 15 per cent for softer material. Frequent anneals may therefore be necessary before the required reduction can be obtained.

Other methods of contracting aluminium include rolling, which is particularly suitable for making shallow grooves, and forming between dies in a press.

### 7.9.3
Curling or beading on circular shells is usually more economically done on a spinning lathe, the press tending to be used for beading operations only on irregular shells or where quantities justify the high cost of press tools.

### 7.9.4
Embossing, coining and stamping are invariably grouped together since each embodies the principle of direct compression of the metal between a punch and a die. Each is used to make shallow impressions, though of different types.

In embossing there is little or no stretching of the metal, the embossed metal being of uniform thickness. The method used to decorate aluminium foil is to pass the foil between a steel roll on which the design has been etched or cut and a smooth rubber-faced roll which applies the pressure.

In coining, extremely high pressures are required to force the metal into the cavity between 'closed' top and bottom dies which, unlike embossing tools, usually carry independent designs; the gauge of the sheet must therefore be closely controlled to obtain consistent results. The aluminium alloy used in this operation should preferably be of high elongation and wide plastic range. Special short-stroke knuckle-joint presses which have the advantage of developing intense pressure towards the end of the stroke are generally used for embossing and coining on a large scale.

Stamping, in this context, refers to the pressing of simple designs of sharp outline, usually trade marks or numerals, on plain surfaces, the impression being on one side only.

### 7.10   JOINING [10]

Connections in aluminium structures are made by riveting, bolting and welding. For a joint to be efficient it must achieve strength properties equivalent to that of the alloy sections or plates to be joined with a minimum number of rivets, bolts or runs of weld metal. The joint should be capable of being made rapidly and effectively either on site or in the workshop, as the case may be, so that the structural members are economically built up and assembled to the required design. Furthermore, the durability of the joints and fixings should not be less than the structural sections joined.

### 7.10.1 Riveted joints [10]

Riveting is still a widely used method of assembling aluminium structures and there is also considerable experience of the use of small tubular rivets up to 9.5 mm diameter in aircraft construction where relatively light gauge sheet and sections are used—see BS 1473 Rivet, bolt and Screw Stock.

Aluminium rivets have generally to be driven cold, and originally the power required to close large rivets is much greater than that for closing steel rivets driven hot. For aircraft and some other structures the high strengths of heat treated alloys are required but for ships and other assemblies exposed to marine environments the non-heat treatable alloy, 5154, is generally used. Before the welding of structural items was practicable it was necessary to hot drive large rivets in this alloy and special portable electric furnaces were developed for this purpose, the rivets being heated to 400–450°C.

Investigations into the cold driving of aluminium rivets with various shapes of heads (or points) has enabled the size of rivets which can be driven cold to be increased—see BS 1974, Large Aluminium Rivets ½–1 in nominal diameters. Suitable alloys are 6082 TB and TF and 2014A TB. The shape of the rivet heads recommended in this specification, although smaller than corresponding heads of steel rivets, is such that when tested in tension the shank fails before the head is torn off. When aluminium rivets are used, there is appreciable elastic recovery after cold driving so that there is little, if any, clamping effect of the rivets on the plates and tension plays only a small part in sustaining the working loads. Further, owing to the smoothness of the jointed aluminium surfaces, the frictional resistance at a joint is only about one-tenth that of similar joints between steel structural members so that almost from the onset the load is taken by shear in the rivets and not by friction in the plates. However, provided the cold driven rivet fills the hole, elastic recovery of the riveted material will exert a grip on the rivets.

Cold driving of rivets in alloys 6082 and 2014A can be made easier by using rivets in the freshly quenched solution-treated condition and cold driving before the rivets had been allowed to age-harden appreciably, i.e. within about 2 h. This time can be considerably increased by keeping the rivets at a low temperature in a refrigerator, e.g. up to a period of 48 h at 0 to −5°C, or six days at −15 to −20°C.

Aluminium alloys cold driven are subject to work hardening so that pneumatic riveters of sufficient power should be used to ensure that the rivets are closed with as few blows as possible. Special rivet points are necessary for larger diameter rivets, and squeeze riveting should be used whenever possible. Rivet spacing should not be less than three times the rivet diameter, and the distance from a cut edge not less than two diameters from the centre of a row of rivets, and an extruded edge not less than one and a half times the rivet diameter.

### 7.10.2   Bolted connections [10]
For connecting aluminium alloy structural assemblies, stainless steel, mild steel (galvanised, sherardised or equally protected) and aluminium bolts are used. Aluminium bolts may be machined from bar in several alloys but for economy in production the heads are usually forged. The alloys used are 2014A, 5056A, 6061 and 6082, as in BS 1473 Rivet, Bolt and Screw Stock.

In bolted connections the same minimum pitches and edge distances as for rivets should be adhered to. As far as possible, bolts should not be used in tension. The use of washers of the same material as the bolt is recommended. Under reverse loading conditions or when vibrations occur, locknuts, spring washers and any alternative device for preventing nuts from working loose should be used.

### 7.10.3   Adhesives [11]
There are many examples of the use of synthetic resin adhesives for making joints between aluminium and its alloys or for bonding aluminium to non-metallic materials, such as plywood, fibreboard, laminated plastics, rubber, cork and glass. Good joint design and meticulous surface preparation are essential where structural strength is important. Any tendency for 'peeling' must be avoided, otherwise the concentration of stress on a very small area of adhesive causes failure at very low stress. Adhesives requiring a low temperature for curing are advantageous when joining heat treatable material. The strength properties of loaded joints depend on temperature and are suitable only for use in the temperature range $-50$ to $100°C$.

### 7.10.4   Welding [12]
Aluminium and its alloys can be readily welded (Table 43). The welding processes fall into three groups, fusion welding, resistance welding and pressure welding, the fusion methods including inert gas shielded welding (Metal Inert Gas and Tungsten Inert Gas) and high energy methods using electron beams and lasers.

#### 7.10.4.1   Oxy-acetylene gas welding
Oxy-acetylene and other gas welding falls in the first group. Before welding, all dirt and grease should be removed by use of a solvent and brushing with a stainless steel wire brush. To get rid of the oxide on the work piece and the filler rod, fluxes are used, their action being both physical and chemical [12]. These are based on alkali fluorides and chlorides. The melting point of the flux must be lower than that of any eutectic present in the product being

Sec. 7.10] **Joining [10]** 225

**Table 43** — Typical processing characteristics of wrought alloys

*(a) Work hardening alloys*

| Material designation | Formability | | Machinability | | Suitability for welding | | | Anodising |
| | | | | | Inert gas shielded arc | Oxy-gas | Resistance spot, seam, etc. | |
| Temper | O | H4 | O | H4 | | | | |
|---|---|---|---|---|---|---|---|---|
| 1080A | E | — | P | — | V | V | G | E |
| 1050A | E | G | F | F | V | V | V | E |
| 1200 | E | V | F | F | V | V | V | V |
| 3103 | E | V | F | F | V | V | E | G |
| 3105 | V | V | — | — | V | V | V | G |
| 5005 | V | G | F | G | E | — | E | E |
| 5251 | V | G | G | G | V | V | E | V |
| 5454 | V | G | G | V | E | F | E | V |
| 5154A | V | F | G | V | E | F | E | V |
| 5083 | G | F | G | V | E | F | E | V |

*(b) Heat treatable alloys*

| Material designation | Formability | | Machinability | | Suitability for welding | | | | | | Anodising |
| | | | | | Inert gas shielded arc | | Oxy-gas | | Resistance spot, seam, etc. | | |
| Temper | TB | TF | TB | TF | TB | TF | TB | TF | TB | TF | |
|---|---|---|---|---|---|---|---|---|---|---|---|
| 6063 | V | G | G | G | V | V | F | F | V | V | V |
| 6061 | V | G | G | G | V | V | F | F | V | V | G |
| 6082 | G | G | G | V | V | V | F | F | V | V | G |
| 2011 | — | F | — | E | — | N | — | N | — | N | F |
| 2014A | G | F | G | V | N | N | N | N | E | E | F |
| 2024 | G | G† | G | V† | N | N† | N | N† | E | E | F |
| 2618A | — | — | — | G | — | N | — | N | — | — | F |
| 7020 | — | — | F | G | V | V | — | — | — | — | F |
| 7075 | — | F | — | V | — | N | — | N | — | V | F |

E = Excellent.   V = Very good.   G = Good.   F = Fair.   U = Unsuitable.   N = Not recommended.
† TD temper

welded and most have a powerful agglomerating effect on metal droplets and rapidly absorb oxides mainly by surface tensions effects. Its fluidity must be high so that it quickly floats to the surface of the pool of molten metal. After welding, the flux must be removed by the use of hot water and scrubbing in order to avoid corrosion in service.

Filler rods are generally of the same alloy as the work, but may be of 5 or 10 per cent silicon alloy. They are coated with flux which must be dry to avoid introducing hydrogen and oxygen into the molten weld metal.

The welding of castings can be carried out with the same facility as welding pure aluminium sheet and plate. However, since castings are less ductile than sheet, preheating of all—or at least part—of the casting is very desirable followed by slow cooling after welding. In the case of sheet and sections it is necessary to take precautions for eliminating the effects of expansion and contraction. Skilled operatives are essential for oxy-gas welding of aluminium and this is mainly used today for repair work.

Heat treated alloys, especially the 2XXX series, are not suitable for joining by oxy-acetylene welding owing to the loss of strength and risk of cracking.

### 7.10.4.2  Metal arc welding

Metal arc welding with flux coated electrodes has been used to a limited extent but is not able to meet current requirements, as the welds tend to be porous and of low efficiency. The MMA (Manual Metal Arc) process has been supplanted by the inert gas processes using, in the UK, argon to shield the weld area.

### 7.10.4.3  Inert gas arc welding [13]

In the tungsten inert gas (TIG) and metal inert gas (MIG) welding processes, a stream of dry argon (99.5 per cent pure) protects the aluminium base metal from reacting with atmospheric oxygen, nitrogen or hydrogen. The argon is fed through an annulus surrounding the electrode and the arc is maintained within this shield between the electrode tip and the work. Fluxes are unnecessary because any oxide on the surface is removed by thermionic emission from surfaces of the aluminium workpiece. In the TIG process, a tungsten or tungsten–zirconium electrode is used with an a.c. electric supply rather than d.c. This is because the action of the arc in removing the oxide film from the workpiece depends on the emission of electrons from the surface when it is connected to the negative pole. When the tungsten is positive it is heated rapidly by the electrons from the aluminium, with the result that the arc is unstable and control of the weld pool is difficult. With an a.c. supply the workpiece is suitably heated during the aluminium-positive half-cycle and the oxide is removed satisfactorily during the tungsten-positive half-cycle. An ancillary high frequency impulse is often superimposed to maintain the arc satisfactorily particularly when starting up. Filler metal is added from a rod of the appropriate composition melted in the arc as it is moved along. For material under about 6.5 mm thickness it is not necessary to bevel the edges before welding, as the penetration of the arc is sufficient without such treatment, but above this thickness they should be bevelled. However, the MIG process is preferred for thicker material, and is shown schematically in Fig. 77.

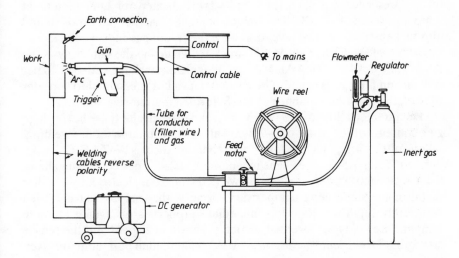

Fig. 77 — Schematic diagram of metal arc inert gas shielded welding process (MIG).

When the tungsten electrode is replaced by a filler wire, the arc is struck between the tip of the wire and the work, with the wire positive because then it fuses rapidly. The work is negative and remains clean. The consumable electrode, MIG process thus offers many inherent advantages including clean welds, arc stability (d.c. supply), full visibility, high welding and deposition speeds which reduce distortion and improve efficiency, little or no touching up, minimum alterations to mechanical properties and corrosion resistance, and reduced residual stresses (promoting improved stress corrosion behaviour). Materials thinner than 3 mm are generally welded by the TIG process, but the 'short arc' MIG process with 0.5 mm diameter wire is suitable for welding sheets 1–6 mm thick. The MIG process is readily automated for repetitive work.

The argon arc welding of the 2XXX series of alloys in the heat treated condition cannot be recommended. The cast weld metal has low properties which are not restored by subsequent working. A weld worked by peening, followed by heat treatment, undergoes structural changes including coagulation of the $CuAl_2$ particles, accompanied by a drastic reduction in properties and the formation of harmful eutectics locally which cannot be redissolved by subsequent heat treatment.

In the welding of Al–Mg alloys of the 5XXX series, a silicon alloy filler is unsuitable and Al–Mg fillers are recommended. The magnesium content of the filler rods should be higher than the parent alloy to reduce the risk of cracking but there may be some susceptibility to intergranular corrosion due to the presence of β particles if the work is heated for long periods at temperatures of 70–200°C. The sodium content of Al–Mg filler rods should also be kept at a very low level to reduce the risk of gassing and cracks.

Alloys of the 6XXX Al–Mg–Si series are welded in the same way as the Al–Mg alloys, often using Al–Mg filler alloy. Braze welding with 5 or 10 per cent silicon filler rods can also be carried out, but it is essential to consider the possible harmful influence of $Mg_2Si$ formed in excessive amounts.

Heat treated alloys of the 7XXX type, and in particular those having high zinc contents of the order of 5–7 per cent, are not suitable for arc welding. Alloys of this series having lower zinc contents such as A–ZAG (2.75–3.5 per cent zinc, 1.5–2.5 per cent magnesium, 0.2–0.7 per cent manganese, 0.1–0.4 per cent chromium) can be welded (tungsten electrode) with filler metal of the same alloy composition, because the cast deposit is cooled sufficiently rapidly to develop immediate properties on ageing at room temperature. A typical welded section is shown in Fig. 78 and the tensile strengths of representative argon arc welded aluminium butt joints are given in Table 44.

### 7.10.4.4   *High energy welding* [14]
The electron beam welding process and the laser technique have both been applied successfully to aluminium-base materials, though few details have yet been published. The fusion welds and heat affected zones are very narrow and properties are good.

### 7.10.4.5   *Resistance welding* [15]
Resistance welding includes the processes of spot, seam and butt welding which make use of the heat generated by an electric current as it passes across the interface between the components to be joined, the weld being made under a suitable applied pressure. Except for the removal of oil and dirt, pure aluminium need not be specially cleaned. For heat treated alloys and those which acquire a high resistance oxide film it is necessary to clean their surfaces where the electrodes make contact.

Most of the wrought alloys can be resistance welded. The high electrical conductivity of aluminium and its alloys makes it essential for the current to be much higher than in the case of steel while, owing to the high thermal

Fig. 78 — Photomicrograph of typical fusion weld in rolled aluminium—4% magnesium plate, showing the weld metal on the left. × 100.

conductivity of the metal, the time of application of the current must be extremely short (e.g. 10 cycles = 1/5 s). A nugget of molten metal is formed within a very short time interval (for instance, a single cycle 1/50 s): this is then work hardened by compression between the electrodes between which the components are held. The pressure is held for a further short period.

For spot welding, the contact points of the electrodes are usually either tipped with a copper–tungsten alloy or may be chromium plated to reduce the tendency to alloy with the aluminium workpiece which causes sticking and increases the temperature of the tip.

### 7.10.4.6 Seam welding [16]
Seam welding may be described as a succession of overlapping spot welds between sheets or between sheet and a section. The electrode contacts consist of two rollers, either or both of which are motor driven, while the surface of one is narrower with a rounded or V shaped edge and that of the other is broader and flat. The current is intermittent and the general sequence of operations is similar to that for spot welding. The same conditions apply regarding cleaning of the metal to be welded as for spot welding.

**Table 44** — Average tensile strength of inert gas arc welded aluminium butt joints

| Alloy type | Alloy designation and temper | Filler alloy type | Tensile strength (N/mm²) | | Percentage elongation on 50 mm | Notes |
|---|---|---|---|---|---|---|
| | | | Unwelded | Welded | | |
| *Wrought forms* | | | | | | |
| 99–99.5% aluminium (soft) | 1200 | 1050A/4043A | 70–105 | 76–95 | 30–10 | |
| 1¼% manganese alloy (soft) | 3103–O | 3103/4043A | 90–190 | 90–120 | 20–5 | |
| 0.8% Mn + 0.8 Mg (soft) | 3105–O | 3103/4043A | 110–160 | 90–150 | 20–5 | |
| 2% Mg (soft) | 5251–O | 5554/5056A | 170–200 | 140–155 | 20–5 | |
| 3½% Mg (soft) | 5154A–O | 5154A/5056A | 215–275 | 150–245 | 18–5 | |
| 4½% Mg + Mn + Cr (soft) | 5083–O | 5056A/5556A | 270–350 | 150–270 | 16–5 | |
| 5% Mg + Mn (soft) | 5056A–O | 5056A | 270–350 | 150–275 | | |
| 2½% Mg + Mn + Cr (soft) | 5454–O | 5454/'5056A | 180–250 | 150–230 | 18–5 | 5056 filler preferred |
| Mg₂Si (soft) | 6082–O | 5056/4043 | 150–185 | 135–185 | 15–5 | Reheat treated. |
| Mg₂Si (fully heat treated) | 6082–TF | 5056/4043 | 280–310 | 230–310 | 10–7 | Hot cracking risk |
| | 6061–TF | 5056/4043 | 240–280 | 200–270 | 8–7 | Hot cracking risk |
| 1% Mg + 4½% + Zn (heat treated) | 6063–TF | 5056/4043 | 180–230 | 170–220 | 8–7 | Hot cracking risk |
| | 7020–TF | 5556A | 320–340 | 300–325 | 8–5 | Not reheat treated |
| Duralumin type (annealed) | 2014A–O | 4043/4047 | 200–280 | 180–270 | 2–1 | Hot cracking risk |
| Duralumin type (heat treated) | 2014A–TB | 4043/4047 | 400–500 | 230–280 | 7–3 | Reheat treated |
| *Castings* | | | | | | |
| 5% Si + 2½% Cu | LM4–M | 4043/4047 | 140–160 | 125–155 | 3–2 | |
| | LM4–TF | | 230–280 | 200–270 | 2–1 | Reheat treated |
| 5% Mg | LM5–M | 5556A | 140–170 | | 5–2 | |
| 12% Si | LM6–M | 4047 | 155–200 | 155–190 | 5–2 | |
| 7% Si | LM25–M | 4047 | 130–160 | 130–160 | 3–1 | |
| | LM25–TF | | 230–280 | 200–250 | 2–1 | Reheat treated |

### 7.10.4.7 Resistance butt welding

This is commonly used for joining wire and rod for electrical purposes, and for wire drawing. It can also be used for joining light extruded bars and sections having approximately equal breadth and width dimensions. The mechanical properties of the weld zone can be improved by an upsetting operation following immediately after the welding cycle. This upsetting produces a flash of solidified metal which contains any oxide 'extruded' from the joint and which has to be removed before further processing.

### 7.10.5 Diffusion bonding

Aluminium and most of its alloys can be joined by diffusion bonding both to itself and to dissimilar metals, particularly carbon steels, stainless steel and copper. Solid state diffusion is the basis of the process when two mating surfaces are held together under pressure for sufficient time at a temperature below the melting point of the materials [17]. Clearly the oxide film must be removed, preferably before the surfaces are brought into contact, and the operation usually takes place in vacuum. For joining aluminium to other metals, intermediate thin layers of a third material are often used to facilitate initial bonding and the subsequent diffusion which is essential to securing a satisfactory weld. Suitable conditions can ensure that diffusion bonds are made rapidly but the conditions have to be worked out for each combination of materials involved.

### 7.10.6 Friction welding [18]

Aluminium alloy round bars and some regular sections are amenable to joining by friction welding in which friction is used to heat the ends and which are then forced into close contact. One or both ends are rotated at high speed in contact and when the required temperature is attained a 'push up' pressure is applied. The process has been used to make sound joints of low resistance between the aluminium and copper bars which conduct the high currents to the electrolytic reduction cells of aluminium smelters. Aluminium to steel welds are also made by this process.

### 7.10.7 Brazing [19]

Pure aluminium and some alloys, such as 3103, can be brazed, provided they do not start to melt at temperatures as low as the brazing filler alloys, usually an aluminium–silicon alloy with 5, 7 or 10 per cent silicon. For this reason the usual casting alloys are never brazed, only the aluminium–zinc–magnesium alloy, DTD 5008B being suitable. Brazed assemblies must be well washed to remove flux residues.

### 7.10.8  Soldering [19]

Soldered aluminium joints, unless adequately protected, are not recommended for applications where they are exposed to weathering or to interior applications under damp, humid conditions. Corrosion problems arise because suitable solders set up galvanic attack, while common soldering fluxes are highly corrosive and it is not always practicable to ensure their complete removal after joining.

Soldering, like brazing, depends for its success on breaking down the oxide film and complete 'wetting' or alloying of the surface to be joined.

## 7.11  Machining [20,21]

Although pure aluminium may be machinined satisfactorily if care is taken with the choice of tool, the harder alloys are generally easier to machine. Pure aluminium and the softer alloys generally produce swarf in a continuous ribbon and need tools with large rake angles and chip breakers for best results. Heat treatable alloys such as 6082, 2014A and 7075 machine very well when fully heat treated and with chip breakers produce swarf in short, tight coils.

The higher silicon content Al–Si alloys are more difficult to machine owing to the hard silicon particles which tend to give poor surfaces, unless they are finished by diamond-tipped tools. Special alloys are available for high speed machining of repetition parts, the best known being designated FC1 (2011). It contains 5–6 per cent copper plus 0.2–0.7 per cent each of lead and bismuth. On machining it forms fine chips that can easily be removed, the low melting point Pb–Bi constituent acting as a chip breaker.

### 7.11.1  Turning

For turning, the tool should have a front clearance of 6–8° and a top rate ranging from 25 to 30°—especially useful for high speed steel tools, with 20° maximum for carbide tools and cutting edge angle of 35–50°. The cutting and clearance faces of the tools must be as fine as possible. A lubricant should always be used for finishing cuts. Since the coefficient of expansion of aluminium and its alloys is high it is essential that the work be cooled before finishing to accurate dimensions. Diamond tools are often used to give a high finish on cast items such as pistons.

### 7.11.2  Threading

Screw threads are cut in the usual way. Too fine a thread should not be cut in aluminium and its alloys, and whenever possible the thread should be larger than would be used for the heavy metals.

**7.11.3   Tapping**

Fluted taps are preferable for tapping holes in aluminium. The flutes must be polished to facilitate chip removal. Thread relief must not be used, otherwise material will be torn out when removing the tap, and special cutting oil must be used.

**7.11.4   Drilling**

The type of drill ordinarily used for aluminium is the standard twist type with a helix angle of 42°, a point angle of 120–140° with wide polished flutes. A point angle of 160° may be used to avoid heavy chip formation around holes in soft aluminium but the drill then centres badly.

Aluminium should be drilled at high speed but with only a moderate feed. The cutting speed of the drill should, if possible, always be over 100 m/min. To obtain perfectly smooth walls to close limits of accuracy, subsequent reaming of the holes is in many cases essential.

**7.11.5   Grinding**

Abrasive grinding is carried out using coarse grit (below 50) carborundum or corundum–rubber bonded wheels or bakelite bonded belts. Wheels are dressed with tallow for fettling and with petroleum jelly for finishing operations in order to prevent the adhesion of metal particles without impairing the cutting ability of the wheel.

**7.11.6   Filing**

For filing, single-cut files do not clog as rapidly as the double-cut type.

**7.11.7   Sawing**

A band-saw of the wood working type is excellent for such work as cutting off runners and risers from castings. The teeth should be large and the speed about 180 m/min. Band-saws should be lubricated with oil or tallow.

High speed circular saws are used for cutting extruded sections to length, and large-diameter saws are used for cutting heavy bars and sections and for cutting ingots to length. Travelling circular saws are used for trimming plate.

**7.12   POLISHING**

The methods used for the mechanical polishing of aluminium differ very little from those common with brass, nickel or silverware. The types of mops and grades of composition are selected according to the degree of finish required. Articles made from sheet or circles, spinnings stampings or extruded sections do not require heavy pressure from polishing mops.

**7.12.1  Surface brightening** [22]

Anodic or electrolytic brightening involves dissolving away the natural oxide film, oxidising the metal beneath and then dissolving the oxide simultaneously. The 'Brytal' process employs an alkaline electrolyte 50 g/l $Na_3Po_4$ + 150 g/l $Na_2CO_3$ at 85–95°C. The anode current density is 4–7 A/$dm^2$ and terminal voltage is 12–14 V. It is applied in two intervals of 10 min with an intermediate break using a process timer. The 'Alzac' process uses a fluoborate electrolyte in hydrofluoric acid with additions of sulphuric, chromic or hydrosilicic acid, for polishing, followed by anodising by the sulphuric acid process. The higher the purity, the higher the degree of brightening attained, and alloys of high purity are available to British Standards (designated BTR1, BTR2 and BTR6).

**7.12.1.1  Electropolishing** [22]

Electropolishing is usually applied after mechanical polishing and provides a surface with a much higher specular reflection than can be obtained by mechanical polishing alone. The degree of reflection depends on the alloy, 99.9 per cent aluminium or metal of higher purity being best in this respect.

**7.12.1.2  Chemical polishing** [23]

Chemical processes have tended to replace electropolishing for finishing mass produced components prior to decorative anodising. These use solutions such as concentrated phosphoric–nitric acid (96 per cent $H_3PO_4$, 4 per cent $HNO_3$ and 2 g/kg $CuNo_3$ catalyst), 3 min at 96°C, as in the Alupol process or the more dilute nitric acid–ammonium bifluoride solutions such as the Erftwork process, developed in Germany by the Vereinigte Aluminium-Werke AG.

**7.13  ANODISING** [24]

Anodising is an electrolytic process whereby the natural protective oxide film occurring on the surface of aluminium and its alloys is thickened. Thorough degreasing must precede the oxidising process. The aluminium is made the anode in an electrolytic cell and the cathode is usually lead sheet. When current is passed, instead of oxygen being released at the anode as a gas, it combines with aluminium to form a layer of porous aluminium oxide. Many electrolytes have been patented but the main compositions only are summarised.

**7.13.1  Sulphuric acid anodising**

The more usual electrolytes are based on sulphuric acid, 15–20 wt%, but other acids such as chromic, phosphoric and oxalic acids are also used commercially.

The oxide film is extremely adherent and the original texture of the metal surface, bright or matt, is not altered by anodic treatment, but surface defects such as non-metallic inclusions, scratches, etc., tend to become accentuated. If welded articles are to be anodised, the welding rods should preferably be of the same composition as the workpiece, as alloy filler rods with a high silicon content give dark films.

The characteristics of the film, particularly hardness, abrasion resistance, density, colour and flexibility, vary according to the electrolyte used, the.temperature, the applied voltage and the alloy being treated. The structure of the oxide layer comprises microscopic hexagonal columns, each with a central pore, in honeycomb formation (Fig. 79) The normal film

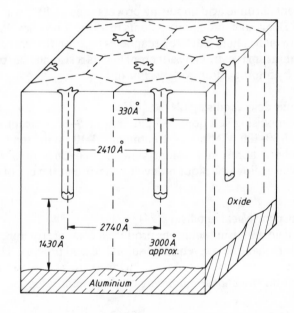

Fig. 79 — The cellular structure of porous oxide films formed by anodising (*after Keller.*)

thickness range is 18–25 $\mu$. The diameter of the pores and the thickness of the barrier layer that is continuously formed during the anodising process between the underlying metal and the growing cells are controlled not only by the particular electrolyte used but also by temperature and applied voltage.

In anodising baths the resistance rises as the thickness of the oxide coating increases. It is therefore unnecessary for the cathodes to conform in shape with the articles connected as anodes in order to produce a uniform thickness of oxide coating. The anode current density is 1.2–1.8 A/dm$^2$ at 10–15 V; the treatment time is 20–60 min; the film formation rate is about 0.3–0.4 $\mu$/min at 1 A/dm$^2$ on 99. 5 per cent Al with a lead cathode.

Normally, sulphuric acid bath temperatures are maintained in the range 18–25°C but by operating between −5 and +5°C, very hard coatings are obtained. When no colouring is required the anodised film is sealed, generally in steam or boiling water for 20 min; this treatment converts the oxide to a hydrated form of larger specific volume. Sealing in aqueous solutions considerably reduces hardness and the film is usually waxed or impregnated with silicone oil.

### 7.13.2 Chromic acid anodising [24]
The 3 per cent chromic acid anodising process used at 40°C was covered by the original Bengough and Stuart patents in 1923. It was developed for the protection of aluminium parts of seaplanes against the corrosion of sea water. The thin film is extremely adherent and its resistance can be enhanced by 'sealing' by impregnation with wax or oil.

### 7.13.3 Oxalic acid anodising [24]
Oxalic acid anodising solutions 3–5 per cent or 5–10 per cent operated at 15–40°C and 30–100 V, time of treatment 10–100 min, tend to produce translucent hard abrasion resistant yellowish coatings of 0.001–0.1 mm thickness. This yellowish colour is developed without the use of added dyes or pigments.

### 7.13.4 Phosphoric acid anodising [24]
Phosphoric acid anodising solutions produce a coating structure with larger pore diameters than the conventional sulphuric acid process. The structure is ideally suited as pretreatment for the electroplating of aluminium and as a pretreatment for the adhesive bonding of aluminium components.

### 7.14 COLOURING

A freshly produced anodic film, particularly when made by the sulphuric acid process, has a porous absorptive structure that readily accepts dye stuffs. Where colour is called for, the anodised, washed and rinsed component is immersed in an aqueous medium containing either an organic dye or a pigment, prior to sealing in steam or boiling water.

In the case of organic dye stuffs the alumina acts as a mordant for the dye and gives good light fastness (of the same order as for fabrics). Inorganic colours are sometimes produced by reactions in the pores of the film, for example, by impregnating the freshly formed film with a solution of a salt followed by immersion in a second chemical which reacts to give a more permanent shade.

An important development was the production of a range of light fast colours by various combinations of metal composition and anodising conditions. These are covered by patents and the semis are known by trade names such as 'Alcancolor' and 'Kalcolor'. They are normally rather dark tones of bronze, grey, reddish browns and gold which are very suitable for cladding panels of buildings with a film thickness normally of 25 $\mu$m but may be up to 30 $\mu$m.

Mention must be made of hard anodising applied to parts which have to withstand wear. The process variables are selected to suit the parent metal which should have sufficient strength to withstand surface pressures in service and be smooth to reduce the risk of abrasion. The characteristics of anodised aluminium are covered in two Standards, namely BS 1615:1972 and BS 3987:1974.

## 7.15  ELECTROPLATING [25]

The electroplating of aluminium is not easy, and plated films may accelerate corrosion in service. For plating, the surface is first pickled, polished and degreased by solvents and electrolysis. The first deposition at 20°C in a bath containing (per litre of water) 300 g NaOH, 75 g ZnO, 6 g CuCN and 170 g KCN produces a friable deposit of $Al_xZn_y$ + Zn. The Zn is removed by controlled solution at 20°C in a mixture of 3 vols $HNO_3$, 1 vol 50 per cent HF and 4 vols water, without attacking the $Al_xZn_y$. The work is returned to the first bath, where Zn is again deposited, this time in a non-friable form. Electroplating is now carried out as if the work was solid zinc. It is washed and immediately immersed in a nickel plating bath to deposit at least 20 $\mu$ thickness of dense nickel plate which may then be chromium plated.

## 7.16  PAINTING [26]

Degreasing alone is not sufficient to provide an effective base for paint on aluminium, but excellent adhesion is obtained on a properly anodised surface. On plain metal the best results are given by degreasing, followed by a pretreatment to provide a chemical conversion coating of the acid–chromate–fluoride or acid–chromate–fluoride–phosphate type followed by washing, drying and immediate painting. Etch primers have been developed containing an acid which provides a suitable surface to which the priming coat adheres strongly. For best results, especially for severe conditions, this can be followed by a zinc chromate coating and then by any suitable undercoats and top coats. Paints or marine anti-fouling compositions containing lead, copper or mercury should never be used, as when any damage to the coating occurs, severe galvanic attack results. Aluminium paint provides excellent top coat resistance for long periods.

## REFERENCES

[1]–[9]  For example: ADA Information Bulletins Nos 9–12:

No.  9—Spinning and Panel-beating of Aluminium Alloys.
No. 10—Deep Drawing and Pressing of Aluminium and its Alloys.
No. 11—Forming of Aluminium Alloys by the Rubber Die Press.
No. 12—Forming of Aluminium and its Alloys by the Drop Stamp.

and Hinxman, H., The Forming of Aluminium Sheet, Sheet Metal Industries, Oct, Nov, 1953; Jan, Feb, March, May, July, Aug, Oct, Dec 1954, January 1955.

[10]  See British Standard Code of Practice, CP 118: 1969, *The Structural Use of Aluminium*, 186 pp. New Edition BS CP 8118 to be published in 1987, British Standards Institution.

Also Brimelow, E. I., *Aluminium in Building*, pp. 209–220 and *Aluminium in Building*, Aluminium Development Association (ADA) Symposium, July 1959, 256 pp.

Symposium Proceedings. *Aluminium in Structural Engineering*, London, 1963, A4, 334 pp., 18 papers and Discussion, organised by Institution of Structural Engineers and the Aluminium Federation, and

ADA Brochures Nos. AB8, 9, 12, 13 and 14:

AB8 —Aluminium and its Alloys in Building—an Introductory Survey.
AB9 —Fully Supported Aluminium Roof Covering—with Notes on Installation Practice.
AB12—Aluminium Windows.
AB13—Aluminium Rainwater Goods.
AB14—Aluminium Corrugated and Troughed Sheeting.

[11]  For example: Brimelow, E. I., *ibid.*, p. 298.

[12]  ADA Information Bulletins Nos. 5, 6 and 19:

No. 5—The Gas Welding of Aluminium.
No. 6—Resistance Welding of Wrought Aluminium Alloys.
No. 19—The Arc Welding of Aluminium.

Van Lancker, M., *ibid.*, pp. 283–338.

West, E. G., *Welding of Non-ferrous Metals*, 1951, Chapman & Hall, London, pp. 129–254.

*Symposium on Welding and Riveting Larger Aluminium Structures*, ADA 1952, Aluminium Federation.

*Proceedings of the Select Conference of Weldable Al–Zn–Mg Alloys,*
Sept. 1969, The Welding Institute, A4, 173 pp., 18 papers.

[13] *Welding of Aluminium and Aluminium Alloys,* British Common-
wealth Welding Conference, organised by the Institute of Welding,
June 1957, 84 pp.

[14] Conference Proceedings, *Advances in Welding Processes,* 1978, The
Welding Institute, 388 pp., A4, 38 papers. (See especially papers 1, 5,
15, 17, 30, 46 and 47.)

[15] ADA Information Bulletin No. 6—Resistance Welding of Wrought
Aluminium Alloys, 60 pp.

[16] As Reference [15].

[17] Bartle, P. M., An introduction to diffusion bonding, *Metal Construction and British Welding Journal,* 1969, (**5**), 1, p. 241.

Barta, I. M., Low temperature diffusion bonding of aluminium alloys,
*Welding Journal,* 1964, **43** (6), pp. 241–247s.

Elliott, S. and Wallach, E. R., Joining aluminium to steel, Part 1:
Diffusion bonding, *Metal Construction,* 1981, **13** (3), pp. 167–171.

[18] Part 2: Friction welding, *Metal Construction,* 1981, **13** (4), pp.
221–225.

Scott, M. H. and Squires, I. F., Metallurgical examination of aluminium–stainless steel friction welds, *British Welding Journal,* 1966,
March.

Duffin, F. D. and Bahrani, A. S., The mechanics of friction welding,
*Proceedings of 3rd International Conference on Advances in Welding
Processes,* 1974, UK, p. 34.

Jessop, T. J., Nicholas, E. D. and Dinsdale, W. O., Friction welding
dissimilar metals, *Advances in Welding Processes Conference,* 1978, p.
49.

[19] *The Properties of Aluminium and its Alloys,* Aluminium Federation
(ALFED), p. 72, 1984.

[20] ADA Information Bulletin No. 7—Machining Aluminium, 56 pp.

[21] Van Lancker, M., *ibid.,* pp. 268–282.

[22] Van Lancker, M., *ibid.,* pp. 354–7.

ADA Information Bulletin No. 13—Surface Finishing of Aluminium
and its Alloys, 44 pp.

[23] Brimelow, E. I., *ibid.,* p. 333.

[24] ADA Information Bulletin No. 14—Anodic Oxidation of Aluminium
and its Alloys, 184 pp.

*The Practical Anodising of Aluminium,* Dr W. W. E. Hubner and
Dipl. Ing A. Schiltknecht, Translated by Winifred Lewis, 1960, Macdonald & Evans, 334 pp.

Lane, J., *Metals and Materials,* March 1986, p. 157.

BS 1615—Anodic Oxidation Coatings on Aluminium—1972.

BS 3987—Anodic Oxide Coatings on Wrought Aluminium for External Architectural Applications—1974.

[25] Van Lancker, M., *ibid.,* pp. 359–363.

ADA Information Bulletin No. 14—*ibid.*

[26 *The Properties of Aluminium and its Alloys,* p. 78, ALFED, 1984.


## FURTHER READING

High speed machining of aluminium alloy castings, *Machinery,* Sept. 1953.

Wernick, S. and Pinner, R., Surface Treatment and Finishing of Light Metals, Sheet Metal Industries (Series of Articles from December 1948—1955).

*Das Chemische Verhalten Von Aluminium, Erfahrungen und Ergenbisse aus Forschung und Praxis,* Aluminium Verlag, 1955, 333 pp.

*Aluminium Workshop Practice,* W. Hegmann, Translated by B. Harocopos (2nd English Edition), Aluminium-Verlag GMBH, 1982, 230 pp.

*The Technology of Aluminium and its Light Alloys,* Dr Ing. Alfred Von Zeerleder, Translated from 3rd German edition by J. Juxon Stevens, High Duty Alloys Ltd, 1948, 452 pp.

*Les Applications de l'Aluminium dans les Industries Chimiques et Alimentaires,* Paul Juniere and Mlle Sigwalt, Editions Eyrolles, Paris, 1962, 332 pp.

ADA Research Report Nos. RR3, RR9, RR16, RR21, RR29, 31, 32, 33, 35, 37, 38.

# 8

# Economic factors

## 8.1 PRODUCTION AND CONSUMPTION DATA

As will be seen in Fig. 80, over the period 1920–1940 there was a steady increase in the world's production of primary aluminium. During the period of the Second World War the rate of production was greatly increased to meet the requirements of the fighting forces, and most civilian applications were cut off from supplies of the metal. With the cessation of hostilities intensive successful efforts were devoted to find markets for the increased capacity which had become available. Not only had the capacity of pre-war primary producers been increased but other companies which had operated Government plants under the technical guidance of long established aluminium smelters purchased the Government plants to operate independently and they too looked for outlets for their metal. The result of much r and d (and sales) activity was an increased demand which led to an increase in production capacity, roughly doubling over the period 1950–1960 and again over the period 1960–1970. In the early years of 1970 the rate of increase in terms of tonnes/annum was maintained until around 1975 when there was a fall in demand for aluminium and there was a cutback in production, with pot-lines being closed down during this period of general depression. However, demand increased again and by 1980 world production of primary aluminium reached approximately $16 \times 10^6$ tonnes/annum but levelled off at this figure for a year or two. Estimates of growth rates to 1990 range from 3 to 5 per cent per annum.

In a paper on a mathematical model for planning future requirements [1], an indication is given of the number of variables involved. Attempts to forecast over long periods are liable to serious errors, and when attempting to deal with global requirements the magnitude of the task becomes even more complex when governments take action to protect national interests.

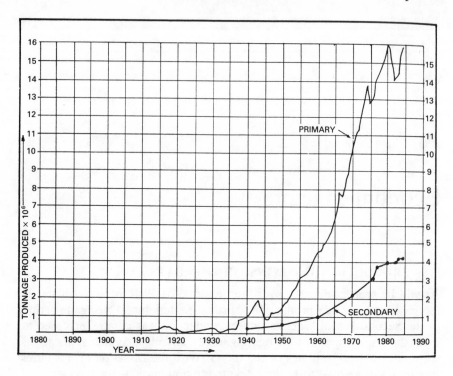

Fig. 80 — World production of primary aluminium 1887–1986. Production of
secondary aluminium in the Western World.

However, the need for forecasting for a period of 5–10 years will be
appreciated, since the time span between planning new mining or reduction
facilities and bringing into production may be as much as seven years.

The production of primary aluminium [2] in 1978 of forty different
countries or areas around the world is given in Table 45. Where available,
figures are also given for the production of secondary aluminium. Prior to
the Second World War, secondary aluminium was a small fraction of the
output of primary metal. The need to conserve metal during the war
demonstrated that reclaimed metal could be used satisfactorily. More
recently it is interesting to speculate to what extent recycling of aluminium
beverage cans has been a significant factor in the acceptance of the metal for
this application [3]. The remelting and use of process discards by semi-
fabricators is normal practice which can be extended to cover major users if
careful segregation into alloy groups is observed. Where alloys are mixed
the segregation into alloy groups becomes too costly and would outweigh
any savings in energy costs compared with primary metal. The metal from
mixed alloy scrap has usually to be used for less critical casting applications.

**Table 45** [2] — Aluminium production and consumption 1985

| Country | Primary aluminium | Thousand metric tonnes<br>Secondary metal | Con–sumption |
|---|---|---|---|
| World total | 16,460 | 4357 | 21,672 |
| Argentina | 140 | 4 | 84 |
| Australia | 852 | 45 | 328 |
| Austria | 94 | 21 | 154 |
| Bahrain | 175 | — | n.a. |
| Belgium (and Luxembourg) | — | 2 | 286 |
| Brazil | 550 | 45 | 411 |
| Cameroons | 82 | — | n.a. |
| Canada | 1282 | 74 | 419 |
| China | 425 | n.a. | 700 |
| Czechoslovakia | 32 | n.a. | n.a. |
| Denmark | — | 16 | 38 |
| Dubai | 155 | n.a. | n.a. |
| Egypt | 178 | n.a. | n.a. |
| Finland | — | 21 | 37 |
| France | 294 | 170 | 742 |
| Germany (Dem. Rep.) | 60 | 65 | 230 |
| Germany (Federal Rep.) | 745 | 457 | 1679 |
| Ghana | 47 | — | n.a. |
| Greece | 123 | — | 98 |
| Hungary | 74 | n.a. | n.a. |
| Iceland | 77 | n.a. | n.a. |
| India | 266 | n.a. | 297 |
| Indonesia | 217 | n.a. | 41 |
| Iran | 43 | n.a. | 2 |
| Italy | 224 | 282 | 833 |
| Japan | 226 | 861 | 2602 |
| Korea (N) | 10 | n.a. | n.a. |
| Korea (S) | 17 | 17 | 163 |
| Malaysia | — | n.a. | 22 |
| Mexico | 43 | 25 | 99 |
| Netherlands | 244 | 62 | 151 |
| New Zealand | 243 | 4 | 39 |
| Norway | 724 | 6 | 139 |
| Poland | 47 | n.a. | 133 |
| Portugal | — | 2 | 38 |
| Romania | 213 | n.a. | n.a. |
| South Africa (and Namibia) | 164 | 30 | 96 |
| Spain | 370 | 55 | 250 |
| Surinam | 29 | — | n.a. |
| Sweden | 83 | 32 | 123 |
| Switzerland | 72 | 26 | 170 |
| Turkey | 54 | n.a. | n.a. |
| United Kingdom | 275 | 158 | 470 |
| USA | 3500 | 1762 | 6162 |
| USSR | 2300 | n.a. | 1850 |
| Venezuela | 402 | 15 | 162 |
| Yugoslavia | 271 | 45 | 213 |

Some of these figures may be subject to correction, as returns are made late by certain countries, and estimates have been made when possible, but others have been marked 'n.a.' indicating 'not available'.

## 8.2   STRUCTURE OF THE INDUSTRY

Bauxite is to be found in many locations around the world and is present in known deposits in quantities to meet requirements up to and including the first quarter of the twentieth century. Exploration continues for new sources of the mineral and research into improvements in processes for the separation of alumina from the ore. Should the need arise, methods are available for the extraction of alumina from clays and from other minerals such as shales.

**Table 46** [2] — Bauxite production in 1975 and 1985

|                     | Thousand metric tonnes | |
|---------------------|-----------:|-----------:|
|                     | 1975       | 1985       |
| World               | 76,885     | 88,019     |
| Australia           | 21,034     | 31,178     |
| Brazil              | 969        | 5486       |
| China               | 1000       | 2100       |
| Dominican Republic  | 785        | —          |
| France              | 2563       | 1530       |
| Ghana               | 321        | 125        |
| Greece              | 3006       | 2367       |
| Guinea              | 8466       | 14,329     |
| Guyana              | 3824       | 2206       |
| Haiti               | 522        | —          |
| Hungary             | 2890       | 2691       |
| India               | 1094       | 2121       |
| Indonesia           | 993        | 831        |
| Italy               | 32         | —          |
| Jamaica             | 11,157     | 6289       |
| Malaysia            | 704        | 492        |
| Mozambique          | 5          | 5          |
| Romania             | 779        | 500        |
| Sierra Leone        | 716        | 1141       |
| Spain               | 6          | 7          |
| Surinam             | 4749       | 3738       |
| Turkey              | 558        | 214        |
| USA                 | 1800       | 674        |
| USSR[†]             | 6600       | 6400       |
| Yugoslavia          | 2306       | 3250       |

[†] Including other aluminium ores (alunite and nepheline).

In 1978 the world total of bauxite production [4] amounted to $83.88 \times 10^6$ metric tonnes from twenty-six countries or areas as shown in Table 46.

In some areas, such as Jamaica where operations are in direct partnership with government agencies, bauxite is mined and refined into alumina for shipping to Canada [5] and elsewhere for reduction to aluminium. Alumina plants in Quebec obtain bauxite from Alcan Group sources and third parties. The world's largest alumina plant is located in Queensland, Australia and its ownership is shared by a number of companies. The bauxite operations in Guinea in West Africa and at Trombetas in the Amazon region of Brazil are also jointly owned, the ore being shipped to participants for refining.

In addition to alumina-refining plants fully owned by individual reduction plant operators the 70s and 80s have been the construction of 800,000 tonne plants at San Ciprian NW Spain and at Aughinish in the Republic of Ireland owned by a number of participants.

It will be appreciated that owing to the widespread nature of the bauxite mining and refining operations, the number of different nations and currencies involved with constantly changing exchange rates, different labour rates and lack of published information, it is not possible to give a typical cost of production of alumina. The position is further complicated in the case of recent installations where recession has necessitated delays in construction programmes and inflation has increased costs.

To ensure some certainty in demand on the one hand and supply on the other, some long term contracts covering a span of 20–25 years may be entered into. In other contracts an exchange of aluminium for alumina may be arranged.

## 8.3  PRODUCTION OF PRIMARY ALUMINIUM [6]

During the early years of the development of the aluminium industry, in N. America production was mainly by ALCOA in the USA. In 1922, ALCAN was formed as a separate company operating in Canada and distributing metal worldwide. For a number of years the two companies were involved in legal actions taken to establish whether or not there had been any breach of the Sherman Anti-Trust Act of 1890. The virtual monopoly of ALCOA in the US continued until 1941 when Reynolds Metal entered the business and they were joined in competition by Kaiser Aluminium when plants built by the US government were sold. Before 1940 the main producers in Europe were Pechiney in France and Alusuisse in Switzerland. The British Aluminium company operated plants in Scotland from 1896, but hydroelectric resources were insufficient to permit the UK to become a low cost major producer of primary metal.

In Europe the stimulus for new aluminium production came mostly from national governments who in the 1960s encouraged the development of local production capacity either alone, or in a joint venture with one of the 'big six' (ALCOA, ALCAN, Reynolds, Kaiser, Pechiney and Alusuisse). Increase in capacity in the UK was essentially due to the support of the British Government in the early 1970s, plants being erected at Anglesey supplied with power mainly from the National Grid, Inverness supplied by hydroelectric power and Lynemouth with its coal-fired generating station supplied from local mines. Production in the UK is now centred at Lynemouth.

### 8.3.1 Energy requirements
The total amount of energy required in the steps leading from the bauxite mine to rolling mill can be broken into two major components: electricity and fuel (or fuel-related materials).

In 1982, for the industry-wide weighted average of 17 cents per kWh, the electricity cost is about $300 per tonne of metal. The fuel bill completely, apart from electricity, also comes to about $300 per tonne of metal. This total energy bill of approximately $600 represents about half the production cost of that tonne on typical existing facilities. At some smelters the cost of power is much higher and can bring the total energy bill to more than $900 per tonne of aluminium. In the early 1970s the industry's total energy bill came to about one-fifth of the overall production cost.

While the worldwide weighted average power consumption is 17,000 kWh per tonne, some modern plants require as little as 13,500 kWh per tonne. Any quantum improvement will need to await the successful development of a non-electrolytic smelting process which does not corrode the process equipment (cf. the subhalide process).

Some 30 per cent of the fuel consumption, in the production of aluminium, is used to produce process steam for which coal-fired equipment is well suited. Just over half the industry's power supply is hydroelectric, just over a quarter is coal based and about 20 per cent comes from oil, gas and nuclear plants. The industry owns and operates about 27 per cent of its power supply and buys the remaining 73 per cent.

Hydroelectricity is attractive to the aluminium industry. It is costly to develop initially but its subsequent economic operation is less affected by rising labour, fuel and operational costs, and is thus a strong hedge against inflation. Equally it is of great interest to those countries with hydro resources whose potential cannot be harnessed unless a large load is introduced.

It is widely believed that the current modest aluminium production based on the vast amounts of natural gas available for power generation in the Middle East and elsewhere is likely to grow, as also will the use of coal-based power.

## 8.4   PRICE OF PRIMARY ALUMINIUM [6]

Several producers publish a metal price list with extras for various shapes, such as ingots for sheet or extrusion, and in the US this is normal. In general there is no variation between the list prices of major suppliers, but producers do not always sell at published prices. When there is an over-supply in the market, producers discount their own list prices to be more competitive. In the USA, discounts are usually announced in the press but this is not an invariable rule. When metal is in short supply some producers seek ways to sell at least some of their metal at a premium.

In Europe it is not common practice to publish metal price lists, and the pricing structure for 'extras' varies from country to country. There is a further complication in the currency fluctuations which occur, often at short intervals, to affect the real prices paid for ore, alumina, primary metal and scrap. From time to time some national governments impose controls on producers' prices and these controls frequently reflect political or social interests or a need to obtain hard currencies rather than commercial considerations.

For European subsidiaries of American companies, the ALCAN export price reported as the 'World price' is frequently cited as a 'reference price' in contracts.

The price at which primary metal is transferred by producers to their own subsidiaries or divisions is never publicised. In determining the 'transfer price' the multinational companies have to consider what effects their pricing policy may have not only on the countries directly concerned but also for the company as a whole.

Merchants, other than primary producers, trade in aluminium as in other metals. Their price depends on supply and demand, and even more on how the trend of the market is perceived by the merchants. The 'free market' or 'spot price' is generally reported in specialist magazines.

The introduction of aluminium into the contracts of the London Metal Exchange (LME) in 1978 made aluminium trading more accessible to countries which previously had difficulty in channelling their output in Europe. Another trading organisation affecting the prices of aluminium is COMEX in New York.

Scrap and recycled aluminium, much of which is used for castings with some in ingots for rolling or extrusion, is yet another variable in the supply-and-demand equation for the aluminium semis industry.

During periods when the 'spot' metal price is lower than producers' prices it is not infrequent for large consumers to buy metal on the free market and have semis producers transform it into the products they want — this is termed 'tolling' and may be considered as a way of eluding a producers price or 'transfer' price. The integrated producers are obviously reluctant to follow that line but in some cases they do.

Many independent extruders have their own melting and casting facilities for producing extrusion ingot from remelted ingot or scrap.

Over the period 1973–1984 the consumption of semi-finished products in Western Europe has increased from around $2.5 \times 10^6$ tonnes to just over $3.0 \times 10^6$ tonnes, falling to as low as $1.7 \times 10^6$ tonnes in 1975 and to $2.7 \times 10^6$ in 1982. During this time the UK (LME) price for primary aluminium fluctuated widely over short periods, the UK-realised price rising from £230 per tonne in 1973 to around £1000 per tonne in 1984. By the end of the eleven years the price increase recovered to a growth line close to that of inflation [3] (see Table 47). A comparison between the production of aluminium and

**Table 47** [3] — Aluminium prices compared to inflation for the years 1973–84

| Year | UK-realised price £ tonne | Inflation (RPI) | Ratio: Realised price inflation (RPI) |
|---|---|---|---|
| 1973 | 225 | 226 | 1.0 |
| 1974 | 296 | 258 | 1.15 |
| 1975 | 342 | 350 | 0.98 |
| 1976 | 388 | 382 | 1.03 |
| 1977 | 618 | 454 | 1.36 |
| 1978 | 590 | 494 | 1.19 |
| 1979 | 710 | 556 | 1.3 |
| 1980 | 775 | 660 | 1.18 |
| 1981 | 663 | 742 | 0.89 |
| 1982 | 620 | 825 | 0.75 |
| 1983 | 760 | 845 | 0.90 |
| 1984 | 975 | 865 | 1.12 |

other metals is shown in Table 48.

As already noted, the availability of vast amounts of energy for the production of alumina from bauxite and for the reduction of alumina to aluminium is an important item in the cost of metal production [8]. Hence,

**Table 48** [2]— World production of primary metals

| | Thousand metric tonnes | | |
| | 1983 | 1984 | 1985 |
| --- | --- | --- | --- |
| Crude steel | 664,000 | 710,000 | 718,000 |
| Aluminium | 14,318 | 15,920 | 16,430 |
| Copper | | | |
|   Refined production | 9646 | 9543 | 9713 |
| Zinc | 6317 | 6592 | 6750 |
| Lead | 5290 | 5427 | 5616 |
| Magnesium | 260 | 329 | 329 |
| Tin | 210 | 208 | 206 |
| Nickel | 687 | 751 | 767 |

energy requirements have an important bearing on where major producers locate their plants. Fig. 9 shows material flows in aluminium production and Table 49 shows the fuel requirements. The energy required for reclaiming

**Table 49** [8] — Fuel requirements for aluminium production per kg hot metal/cast ingot UK smelters (1976 data)

| | Hot metal production | Ingot casting |
| --- | --- | --- |
| Electricity, kWh | 17.41 | 0.061 |
| Natural gas, $Nm^3$ | 0.027 | 0.021 |
| LPG, kg | 0.08 | 0.060 |
| Heavy fuel oil, l | 0.0012 | — |
| Medium fuel oil, l | — | 0.004 |
| Light fuel oil, l | 0.0015 | 0.0043 |
| Gas oil, l | 0.0012 | 0.0016 |
| Lubricating oil, l | 0.0003 | — |
| Petrol, l | 0.00034 | 0.0013 |
| Diesel oil, l | 0.0027 | 0.00038 |

secondary metal varies with the type of scrap and the amount of primary metal used. Hancock [8] quotes figures ranging from 12.9 to 36.79 per cent of those for primary metal.

## 8.5  SEMI-FINISHED PRODUCTS [6]

The aluminium semi-finished product industry may be divided into:

  (i) rolled products (plates, coils, strip, foil, circles and slugs);
 (ii) extruded products (shapes, rod, bars and tubes);
(iii) drawn products (wire and cable);
(iv) forgings.

Drawn products and forgings are special categories representing only 11 per cent of the European semi-finished aluminium consumption.

The rolled product industry is dominated by the 'integrated' companies which are part of the larger companies or groups also producing primary metal, whereas 'independent' companies form a large proportion of the extruded product market. These differences are accounted for by different entry barriers in the two parts of the industry.

The number of semis producers in Europe is over 120 companies. In the rolled products' market it is estimated that Pechiney/Cegedur, Alusuisse and Alcan between them command about 50 per cent of the European market. In 1982 in the rolled products' industry, state-owned companies accounted for about 50 per cent of European production.

## 8.6  SEMIS PRICING [6]

### 8.6.1  USA

In the USA the three integrated producers, Alcoa, Kaiser and Reynolds, have a large share of the American semis market and, because of their dominant position in the industry, these three companies are price leaders.

The prices of semis in the USA are controlled essentially by the counteracting forces of supply and demand, as they are or as they are perceived by the customers and suppliers. If one of the leading producers announces a price increase, the others may or may not follow. If the majority accept the increase it becomes effective through the market, but, if one or more of the leading producers refuse to follow, the company which took the initiative normally withdraws the change. In difficult market situations, it is not unusual for price increases, after being announced, actually to be rescinded and price lists re-issued at old levels. The actual market prices, however, do not always coincide with the list prices; they are frequently discounted, with discounts generally announced through the press.

A change in ingot price does not always cause a corresponding change in the price of semis. The prices of different semis products such as extrusions, sheets, plates, rods and bars may change independently of one another. Commodity products, such as can stock or foil stock, are priced separately and their prices may change independently of other products' prices.

The Sherman Anti-Trust Act (1890) mentioned earlier and the Robinson Patman Act (1936) aim to ensure that all customers, who are competitors among themselves, will be treated equally if the cost of serving them is essentially the same; both have a significant effect on price decisions in the USA. From time to time the American Government seeks to regulate the market by issuing guidelines on pricing.

### 8.6.2 Europe

Price levels vary widely between the different European countries, some of which do not have published price lists. In some cases foreign producers are the only companies which set price levels at least for certain categories of product. Published producer lists are not standardised from country to country and often reflect past local conditions rather than present.

Each European country has different anti-trust legislation; however, all companies which manufacture or sell in the Common Market, even if their headquarters are not within the EEC, are subject to the anti-trust laws of the Treaty of Rome which take precedence over national laws regarding competition. Essentially the Treaty of Rome prohibits any agreement between companies which is likely to affect trade between EEC countries by preventing or restricting competition within the EEC.

Table 50 shows that between 1960 and 1980 while the consumption of

**Table 50** [6] — Consumption and intertrade of aluminium semi-finished products in Europe 1960 and 1980

|  | $10^3$ tonnes | |
|---|---|---|
|  | 1960 | 1980 |
| Consumption | 817 | 3290 |
| Intertrade | 78 | 960 |

aluminium semi-finished products has quadrupled the intertrade within Europe has expanded twelve fold. As imports increase, the importance of the leading semis producers in each country is diminishing and specialisation is becoming more important.

Several — if not most — of the European semis producers have different home and export prices; in their domestic market they are concerned with selling at the market price level, but with exports they frequently look more to the possible 'contribution margin'.

For accounting purposes, certain costs such as rates and equipment depreciation are considered as fixed. Any increase in throughput contributes to a reduction in the total of these costs per tonne for the given accounting period.

**REFERENCES**

[1] Van Gameran, A., *Metals and Materials*, May 1978, p. 42.
[2] *Metal Statistics 1985*, published by Metallgeslschaft AG annually.
[3] Hawkins, J. F., *Metallurgist and Materials Technologist*, August 1984, p. 414.
[4] Whitten, N. and Hoskins, C., *Metals Technology*, **11**, July 1984, p. 300.
[5] Alcan Compass, July/August 1981, p. 14.
[6] Camatini, G., *Metals Technology*, May 1983, **10**, pp. 182–187.
[7] Lester, M., *'Chemea 82': 10th Australian Engineering Conference*, Alcan Compass, Sept/Oct. 1982, p. 19.
[8] Hancock, G. F., *Metals Technology*, **11**, July 1984, p. 290.


**FURTHER READING**

*Dealing on the London Metal Exchange*, 1974, published by M. C. Brackenbury and Co. London, 116 pp.
*World Bureau of Metal Statistics*, monthly data published by Bureau of Metal Statistics, London.
*Metal Bulletin Handbook*, published annually; also *Metal Bulletin* periodicals.
*Statistical Year Book*, published annually, United Nations.
*'Aluminium' — a Metal Bulletin World Survey Special Issue*, December 1963, 202 pp.

# 9

# Applications

Aluminium and its alloys have properties which make them suitable for a wide range of applications. The metal is available in a wide variety of forms ranging from thick (250 mm) plate to thin foil (0.005 mm), heavy duty sections for structural engineering to small mouldings for decorative trim, large-diameter rod and bar down to fine wires. Materials are available in a range of tempers to meet requirements of strength and formability combined with low density, whilst some have good thermal and electrical conductivities. Its ability to resist corrosion and to take a variety of finishes is particularly advantageous.

Aluminium provides a particularly interesting example of the development of usage of a metal becoming available at a late stage in the historical context of materials able to meet the requirements of mankind, as it gives an almost unique record of the effect of changing industrial factors. The older metals, iron and steel, copper, brass and bronze, lead, tin and, of course, gold and silver, have been used for many hundreds of years, changing gradually and being replaced or supplemented from time to time with the new materials of the day. Aluminium has achieved its current diversity of large-scale uses in part by replacing the older established metals and in part by linking up with pioneers of new industries based on new technologies. In its early days aluminium was handicapped by being regarded as a precious metal suitable mainly for statues, objects d'art, jewellery and expensive personal knick-knacks. As its cost fell, its range of possible uses expanded but its potential was not fully appreciated until demands of new inventions made engineers and designers appreciate its combination of properties. This was particularly so when aircraft began their spectacular development in the early days of the twentieth century, when the value of low weight was preeminent but this was also recognised in the same period when the internal combustion engine was making possible the motor car and other road vehicles. It was only in the light of their requirements that the discovery of age-hardening to give strong alloys that aluminium could be exploited.

The increasing demands of aircraft builders and engine makers led to major research projects to further improve aluminium alloys, and, in turn, stronger materials able to withstand higher temperatures resulted in further new applications. One of the earliest observations, namely the ability of aluminium to resist atmospheric corrosion, formed the basis of many uses and in the 1930s the aluminium–magnesium series began the large-scale adoption of these alloys for marine purposes. The versatility of processes applied successfully to aluminium from simple castings by craftsmen to complex extruded sections provided further opportunities for competition with other materials and for developments of new applications in their own right.

Throughout the history of its application, it was apparent that aluminium seemed to have the significant disadvantage of a high price per tonne relative to other metals, particularly iron and steel, but, as metals are used largely on a volume basis, the low density and ease of working assisted their competition. Exploitation of other advantages, such as the ability to produce an anodic film that was protective and could also be employed decoratively, opened up other fields of application. Many additional demands on aluminium were made as a result of the war-time requirements of 1914–18 and 1935–45 but the new knowledge of design and material properties were fully used in the subsequent peace-time developments, particularly since the Second World War — in buildings from modest bungalows to giant multi-storey blocks, from dinghies to large tonnage applications on ships, from the cast aluminium coffee pot of Charles Martin Hall to enormous welded aluminium plants for the chemical industry and today's cryogenic applications.

Throughout the first hundred years of its use as an industrial metal there have been continual changes in the percentages of the annual consumption of metal used in different spheres of application. Whilst aluminium has replaced copper in many electrical applications such as high tension power lines and busbar conductors, it is itself challenged by improvements in steels for lightweight structures or for cryogenic applications such as tanks for the transport of liquefied natural gas, LNG. Synthetic organic films are replacing aluminium foil as a wrapping material for some commodities such as tobacco. In aircraft construction, aluminium alloys have to face the challenge of titanium and its alloys and of fibre reinforced resins. Optical fibre technology using glass is replacing metal conductors in telecommunications.

To meet these challenges the aluminium industry continues to be engaged in research and development of processes such as improving the efficiency of the reduction process and in the conversion of the metal to semi-fabricated forms. The rapid solidification process (RSP) and powder metallurgy processes and aluminium alloy fibres offer possibilities of new fields of applications.

Table 51 gives an indication of the way in which changes in industrial

**Table 51** [4] — UK consumption of rolled aluminium products, 1973 and 1983

| Application tonnes | 1973 × 10³ | Percentage | tonnes | 1983 × 10³ | Percentage |
|---|---|---|---|---|---|
| Transport: | | | | | |
| Road | 44.2 | 17.8 | 29.4 | | 11.4 |
| Aircraft | 6.0 | 3.4 | 7.1 | | 2.8 |
| Other | 16.2 | 6.5 | 8.4 | | 3.3 |
| Engineering: | | | | | |
| General | 32.4 | 13.0 | 45.5 | | 17.5 |
| Electrical | 6.3 | 2.5 | 7.1 | | 2.8 |
| Building | 19.2 | 7.7 | 22.8 | | 8.8 |
| Foil | 46.9 | 18.8 | 45.5 | | 17.6 |
| Other packaging, chemical and food | 36.6 | 14.7 | 68.5 | | 26.6 |
| Holloware | 13.2 | 5.3 | 7.1 | | 2.7 |
| Other domestic and office | 14.4 | 5.8 | 4.8 | | 1.9 |
| Other | 13.8 | 5.5 | 12.0 | | 4.6 |
| Total | 249.2 | 100.0 | 258.2 | | 100.0 |

activity in the UK have resulted in changes in the use of aluminium. The comparative decline in the British car industry and the import of cars reduced the percentage for this application from 17.8 to 11.4 per cent of the rolled products' consumption. There was also a slight fall in the consumption of foil but for other packaging, chemical and food there was an increase from 14.7 to 26.6 per cent. There was also a decrease in the amount of aluminium, used for holloware in competition with imported cast iron and stainless steel.

In Table 52 the consumption of semi-fabricated products is given in kg per capita in 1982. It will be seen that in the USA more than twice the amount per capita is consumed than in the European countries, France, Italy and the UK, and about 50 times that for India. Not only do the amounts differ so widely, the applications in different countries also differ. However,

**Table 52** [5] — Semi-fabricated products: consumption in kg per capita in 1982

| | |
|---|---|
| USA | 16 |
| West Germany | 12.6 |
| Japan | 11.4 |
| Italy | 7.4 |
| UK | 7.1 |
| France | 7 |
| India | 0.3 |

the types of alloy found suitable for different applications are much the same in all countries. Table 53 lists the more commonly used wrought alloys for different applications, and Table 54 uses of analyses and casting alloys in industry. For applications where specific properties such as high specular reflectivity are required, alternatives to the alloys listed may prove to be more suitable and such a choice can be resolved by consultation with the supplier.

Table 55 gives aluminium applications by methods of manufacture.

**Table 54** [14] — Analysis of end uses of aluminium alloy castings

| | Percentage |
|---|---|
| Aircraft | 6 |
| Building | 6 |
| Domestic utensils and appliances | 11 |
| Electrical plant and equipment | 9 |
| Machine tools and other mechanical engineering | 6 |
| Office equipment | 2 |
| Railways | 2 |
| Road vehicles (mainly in i.c. engines) | 50 |
| Ship building and marine engineering | 5 |
| Textile plant | 3 |

There are many examples of beautiful monumental work and statuary which have been executed in aluminium [2], the classical example being Eros (June 1893) surmounting the Shaftesbury Memorial fountain in Piccadilly Circus, London.


## 9.1 ROAD TRANSPORT

The advantages of a high strength/weight ratio have been applied to all forms of transport such as rail coaches, motor cars and trucks, vans and 'buses, as well as major elements of ship structures. Aluminium was

**Table 53** — Some wrought alloys used for different applications

|  |  | Applications | Alloys |
|---|---|---|---|
| Transport | Cars, trucks, vans, 'buses | Floor planks, chassis members, container frames, Bodywork | 2014A, 6061, 6082, 7020, 3103, 5154A |
|  | Aircraft | Structures<br>Engines | 2014A, 2034, 7075, 7079, 7178<br>2018, 2618 |
|  | Marine | Superstructures | 5154A, 5454, 6061 |
| Engineering | General | Welded structures | 5154A |
|  |  | Stressed structures | 2014A, 5083, 6061, 6082 |
|  |  | Rivets | 1050A, 2014A, 5056, 5754A |
|  |  | Bolts and screws | 2014A, 5056A, 6061, 6082 |
|  | Electrical | Conductors | 1350, 5005, 6061, 6101A |
| Building | Structures |  | 2014A, 6061, 6063, 6082 |
|  | Roofing and siding |  | 3103, 3105, 5251 |
|  | Rainwater goods |  | 3105 |
|  | Architectural fittings |  | 6063, 6082 |
|  | Mobile homes |  | 3105 |
| Holloware |  | Pressings and spinnings | 1200, 3103, 5005 |
| Food packaging |  | Foil | 1200 |
|  |  | Beverage cans | 5182 |
| Chemical equipment |  | Storage tanks | 3003, 3004, 5254, 5652 |
|  |  | Pressure vessels | 3003, 5456 |
|  |  | Pressure vessels unfired | 5083 |
| Painted sheet |  |  | 1100, 3003, 3105, 5005, 5050, 5052 |

Fig. 81 — An early twentieth century Rolls-Royce car with aluminium body panels
hand formed and attached to timber (ash) frame.

used in most of the earliest automobiles for bodywork (Figs 81 and 82) as
well as engine components but, with the advent of the mass produced car,
the amount per vehicle has remained relatively static other than for the more
costly vehicles. This may be attributed to a number of factors, particularly
the relatively higher cost compared with steel, the early problems with
welding and with repair in the event of damage. Another significant
consideration as far as car manufacturers are concerned has been that of
availability of material of the required size, quality and quantity from at least
two alternative suppliers.

The rapid expansion of primary aluminium production in the past fifty
years and the installation of rolling facilities capable of handling heavy coils
of wide material in a range of work hardened and heat treatable alloys offer
solutions to the supply problems. Developments in welding and joining
techniques, and of alloys suitable for these methods of assembly, open the
way to an extended use of aluminium for applications in the whole field of
transport. Polymeric materials may present a viable alternative to steel for
body panels and it would appear that in the near future the major increases
in the use of aluminium in cars will be in engine components, wheels, chassis
members and body frames. Over half of the cars produced in Europe already
incorporate aluminium radiators (Fig. 83), replacing copper at lower cost
and useful weight saving. Indeed, any higher cost due to the initial price of
aluminium against cheaper alternatives can be offset by savings in fuel which
could amount to 10 per cent.

By 1983 the amount of aluminium in the average American car had risen
to 61.5 kg and in the UK car it was about 19 kg. In the development of the
energy-efficient car, methods for weight saving are under constant review
but any major change from one type of material to another, involving costly

**Table 55** — Aluminium applications by method of manufacture

| Commodity group | Process | Product | Applications |
|---|---|---|---|
| Unalloyed aluminium | Sheet rolling | Plate and sheet | Chemical engineering |
| | | Corrugated sheet | Tanks, panels, roofing and sidings |
| | | Circles | Holloware |
| | | Impact extrusion slugs | Collapsible tubes for pharmaceuticals etc. |
| | | Container sheet | Containers and closures |
| | | Lithographic sheet | Printing |
| | | Foil | Food packaging, kitchen foil, decorative wraps |
| | Continuous casting, rolling and drawing Drawing | Rods and wire | Electrical conductors, rivets, wire for general purposes |
| | Extrusion | Busbar | Electrical conductors, mouldings, tubing |
| | | Shapes | |
| | | Tube | |
| | Atomisation and attrition | Powder and paste | Ink, paint, explosives, aluminothermic mixtures |
| | Casting | Ingot of various sizes | Deoxidation of steel, alloying with heavy metals |
| Casting alloys | Sand casting | Castings (one upwards) | General engineering and building components (any size) |
| | Pressure die casting | Castings in quantities | Housings and castings for general and automobile purposes |
| | Gravity die casting | | Automobile and diesel pistons, casings, etc. Household products |
| | Ornamental casting (lost wax) | Single (or few) castings | Art metalwork, statuary, architectural features |
| Aluminium–silicon alloys | Alloying and casting | Ingot | Hot dipped coated steel components |
| Complex alloys | Rapid solidification processes | Rapidly solidified products (RSP) | Powder metallurgy |
| Wrought aluminium alloy — plate, sheet and coil | Flattening | Flat sheet and plate | General engineering Panels |
| | Bending and folding | Sections and shapes | Marine engineering Structural engineering |
| | Corrugating | Building sheet | Roofing and sidings |
| | Blanking and pressing | Flat shapes and pressings | Holloware, beverage cans, containers |
| | Welding | Seam welded tube | Irrigation tubes, tanks |
| | | Welded structures | Structural and marine engineering |
| | Solution heat treatment | High strength plate and sheet | Aircraft construction, fighting vehicles, structural engineering |
| Wrought aluminium alloy sections | Extrusions | Rod | machined products |
| | | Bar | Machined products, forgings |
| | | Shapes | Aircraft structures, general engineering |
| | | | Vehicle construction |
| | | | Buildings, doors, window frames, greenhouses |
| | | Hollow shapes | Structural work |
| | | | Refrigeration units |
| | Cold drawing | Seamless tube | Structural engineering |
| | | | Furniture |
| | Wire drawing | Wire | Electrical conductors, rivets, nails, screws |

changes in equipment and procedures, is one which takes several years to implement.

Aluminium alloys have been particularly successful in the commercial road vehicle market where they have replaced steel and timber for bodywork on trucks (Figs 84 and 85) and vans as well as chassis structural items, panels and seats in passenger coaches and 'buses (Figs 86 and 87). Again, savings in weight result in fuel economies and also useful savings in

Fig. 82 — Load-bearing structure of Rover Group 'Concept' car to which aluminium alloy pressings are attached by adhesive bonding. (*Courtesy British Leyland and British Alcan.*)

Fig. 83 — Automotive heat exchangers — brazed aluminium sheet and foil finstock. (*Courtesy British Alcan Sheet.*)

Fig. 84 — An early example of a welded aluminium road tanker. (*Courtesy The Welding Institute.*)

Fig. 85 — Technology demonstration truck with an aluminium chassis and bodywork. (*Courtesy British Leyland and British Alcan.*)

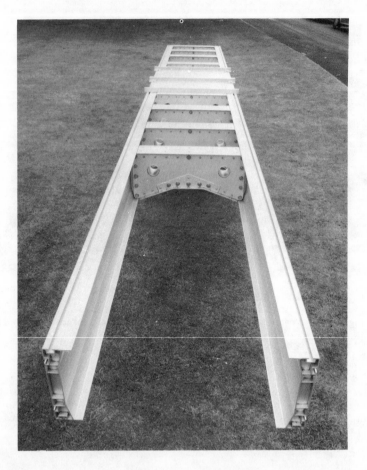

Fig. 86 — An all-aluminium tubular monocoque chassis based on specially designed
extruded sections. (*Courtesy British Alcan.*)

maintenance due to lack of rusting or rotting of timber. A minimum of 10 per
cent saving in fuel is usual and is often twice as much with well designed
vehicles using aluminium alloys properly. For engines the high conductivity
of aluminium is particularly valuable in giving higher efficiency.

A considerable weight of aluminium castings goes into engine blocks,
pistons, cylinder heads, crank cases and gearboxes and it is now possible for
engines to run without steel liners. Road tankers are particularly large users
of aluminium, as are caravans and other types of mobile home.

Fig. 87 — London Transport 'Routemaster' 'bus designed in 1953 as an aluminium monocoque structure of which several thousand have been in continual service since 1954.

## 9.2  RAILWAY VEHICLES

Aluminium was used in Britain for roof panels of railway coaches in the early days of the twentieth century but major applications in this industry were not developed until the early 1950s when both British Rail and particularly London Transport adopted aluminium alloys for their new designs of coaches because the weight savings reduced power demands; for example, London Transport saved at least 12 per cent of the electric power for its aluminium-bodied trains and furthermore reduced maintenance by running these 'Silver Trains' unpainted. The body shell of London Transport coaches exploits to a considerable degree the ease with which complex structural members can be designed for production as extruded sections whilst the familiar sliding doors are cast in an aluminium–silicon alloy. A great many of the internal fittings in British Rail coaches are also in aluminium, as are major components in the diesel powered locomotives (Fig. 88).

## 9.3  AIRCRAFT

By virtue of its low specific gravity and the wide range of mechanical properties of aluminium alloys, stron alloys are widely applied in aircraft construction, accounting for over 75 per cent of the total weight. There were cast aluminium components in the engine which powered the Wright Brothers first successful flight and the first all-aluminium aeroplane was the

Fig. 88 — A modern Tyneside Metro train with aluminium alloy body. (*Courtesy British Alcan.*)

'Silver Streak' in 1921. The then-new aluminium alloys of the 'Duralumin' type also provided the structure for the German 'Zeppelins' of the First World War, and the developments to the modern aircraft led to many new applications in more mundane fields. In the 1950s when it was forecast by many that aircraft speeds would so increase that the aluminium alloys would be unable to withstand stress at the high temperatures resulting from air friction at speeds above Mach 1, it seemed likely that technical developments in that industry would mean less aluminium being used, but improved alloys, allied with engineering design, have permitted Concorde to be constructed largely of aluminium-base materials, even though its speed exceeds twice the speed of sound, Mach. 2.

In the aeroplane of the early 1980s some 75 per cent of the structure is made from established high strength aluminium alloys (Figs 89). At the production peak, consumption was estimated at some 20,000 tonnes of aluminium alloy in the finished aircraft (450 per annum) derived from purchases of, perhaps, five times this quantity. To reduce weight, many aircraft components are machined from thick plate instead of being assembled from sheet and extrusions bolted or riveted together, the savings in operating costs justifying the increased material and construction costs.

The demands of the aircraft industry for still higher combinations of strength, weight and stiffness/weight ratios have stimulated research throughout the world's industrial countries and the latest alloys based on the aluminium–lithium system are in their initial stages of application. They have the benefit of not only satisfying the strength requirements but also have a higher modulus of elasticity permitting stiffer structural elements to be produced with significant savings in weight. This is typical of the way in which development of new applications based on research has been achieved.

Fig. 89 — Production of major aircraft components by milling from solid aluminium alloy plate. (a) High strength aluminium alloy plate being rolled on 3.7 m wide rolling mill at Alcan Plate Ltd.

## 9.4  GENERAL ENGINEERING

The use of high strength heat treatable alloys in general engineering was limited by the general unsuitability of these alloys for joining by welding, and the aircraft industries practice of machining integrally stiffened components from plate does not offer a viable solution. Despite this limitation, the general engineering products which are made from aluminium range from rivets to bridges and television masts. As a specialised high quality product, lithographic sheet is an example of a product which has replaced competitive materials which have been established over many years.

Fig. 89 — Production of major aircraft components by milling from solid aluminium alloy plate. (b) A numerically controlled milling machine cutting out a Jaguar wing skin.

Fig. 89 — Production of major aircraft components by milling from solid aluminium alloy plate. (c) Jaguar fighting 'plane. (*Courtesy British Alcan.*)

## 9.5  AGRICULTURAL ENGINEERING

In agricultural engineering, aluminium alloys are used in the form of corrugated sheet for roofing and sidings of buildings and grain silos, and as extruded or seam welded tube for irrigation purposes where their light weight makes handling and transport much easier. The metal also finds application in farm machinery and for tanks of milk transporters. As extruded sections it is used for the structure of large commercial and garden glass greenhouses, cold frames, etc.

## 9.6  ELECTRICAL ENGINEERING

A considerable tonnage of aluminium is consumed for power transmission cables (Fig. 90), both as ACSR (aluminium stranded steel reinforced) for overhead high tension cables and as aluminium sheathed cables for above ground power distribution. The latter results in a 20–70 per cent saving in weight compared with lead sheathed cable.

The heat treatable Al–Mg–Si 6101A alloy wire has slightly lower conductivity but higher mechanical properties than pure aluminium 1350 and finds use as a low tension power distribution.

Aluminium and its alloys with their natural oxide film of high electrical resistance require special joining techniques such as cold pressure welding.

Fig. 90 — Aluminium power cables.

In relatively small tonnages, aluminium extruded sections have been used as telecommunication wave guides. Aluminium alloys of pressing quality cold rolled sheet are used in place of brass for the manufacture of lamp caps for electrical light bulbs.

Various electrical accessories such as cable boxes, switch boxes, junction boxes, bends and covers have been cast in aluminium alloys.

Because of lower electrical conductivity compared with copper, aluminium conductors require to be of larger cross-sectional area than the equivalent copper conductor. This is a factor which militates against the use of aluminium for house wiring and for the windings of electrical motors or transformers.

In the form of rolled plate sawn to narrow widths, or as extruded shapes, aluminium alloys 1200, 1350 and 6101A are used in many busbar applications.

The non-magnetic characteristics of aluminium make it useful for electrical shielding such as busbar housing and enclosures for other electrical equipment.

## 9.7  MARINE ENGINEERING

The aluminium–magnesium and the aluminium–magnesium silicon alloys, with their combination of good mechanical properties and good corrosion resistance, are suitable for welding and have been used for small craft, superstructures of ocean liners and containers for the transport of liquefied natural gas (LNG). These outlets for rolled and extruded products have decreased owing to a number of reasons but mainly because of the introduction of glass fibre reinforced resins for small craft, the increase in air transport and the development of special steels for low temperature applications.

## 9.8  MILITARY APPLICATIONS

The alloys most favoured for bridge construction, fighting vehicles and armour plate (Fig. 91) are of the aluminium–zinc–magnesium series since they can be welded in thicknesses of up to 50 mm, have low sensitivity to solution heat treatment temperature and have low quench sensitivity coupled with natural ageing properties. They lose little strength once a period of about thirty days have elapsed [1].

## 9.9  BUILDING CONSTRUCTION

After the Second World War, thousands of tons of aluminium alloys reclaimed from wrecked aircraft were used in the form of sheet and extruded sections for the construction of the AIROH house (Fig. 92) to provide temporary housing in and around areas which had been subjected to

Fig. 91 — Welding of aluminium armour of Scorpion tank. (*Courtesy The Welding Institute.*)

Fig. 92 — Assembly of the prefabricated aluminium bungalow 1946.

bombing. These were designed for a life of ten years because the materials used did not have optimum resistance to corrosion but many of the 75,000 built in the former war-time aircraft factories were still occupied after more than a quarter of a century. Many of the lessons learned in designing house components with their successful service led to the confidence of the architects and builders in developing present-day applications.

The prefabricated aluminium bungalows other than the roofing sheets were protected by painting but today curtain wall panels, window frames, sills, doors and their surrounds are usually anodised, as also are decorative features manufactured from wrought products. The use of colour anodised coatings has led to a large variety of finishes available to the architect and coatings of 25 m provide adequate protection even in industrial atmospheres but the anodised surfaces are cleaned periodically. Aluminium is suitable for rainwater goods but not for plumbing or heating services.

Aluminium in the form of cast spandrels, sheet and extruded sections has been used over a great number of years for the facing of buildings. A very important property of aluminium spandrels is their freedom from coloured corrosion products which cause unsightly staining and streakiness on adjacent surfaces of the building. If desired, the spandrels may be anodised and self-coloured [2].

Aluminium alloy sheets of the work hardening type such as 3103 (1.2 per cent Mn) are used extensively for the roofing and siding of buildings, a range of different shapes of corrugation and methods of joining and fixing being enployed. For weatherings and flashings, superpurity aluminium (99.99 per cent) is available; it does not suffer from 'springback' [3].

Building and construction provide the largest outlet for aluminium extrusions. The alloy 6063 is used for thin sections, having an inherent good finish for windows (Fig. 93) and doors. Where greater strength is required the alloy 6061 or 6082 may be used.

When designed to have sufficient stiffness to meet safety requirements, aluminium and its alloys have many applications in the design of stairways, balustrades, escalator casings, lifts, etc.

Castings are used to replace complicated riveted assemblies in aluminium structures. When stronger and tougher components are required and quantities justify the costs involved, it may be preferable to use forgings.

When using aluminium in building it is necessary to avoid corrosion due to contact with alkaline building material such as certain mortars, plasters and cements and to take precautions against galvanic corrosion due to contact with other metals.

Internal fittings such as partitions, doors, venetian blinds, grilles, screens and door furniture are also made in aluminium alloy. Space heating

Fig. 93 — Window frame fabricated from extruded aluminium sections.

radiators for domestic and commercial buildings are now available, built of aluminium extrusions [4]. This application has necessitated a new approach to joining and the inhibition of water corrosion.

Aluminium is used in the manufacture of equipment for building construction such as booms, cranes, hoists, pumps, forms and scaffolding.

## 9.10 STRUCTURAL ENGINEERING

The successful design of major structural components for military aircraft in the Second World War combined with the interest aroused by the aluminium bungalow building programme led to new concepts of structural engineering (Fig. 94) in which full advantage could be taken of the high

Fig. 94 — Triodetic roof structure made from aluminium alloy tubing at Romford swimming pool — also aluminium ventilation ducting. (*Courtesy British Alcan.*)

strength/weight ratio, as exemplified by the construction of very large span buildings such as aircraft hangars. The ease of maintenance and versatility of the extrusion process made the adoption of aluminium to the building of schools and, later, multi-storey buildings acceptable.

For the Festival of Britain Exhibition in 1951 the Dome of Discovery,

111.5 m (365 ft) diameter, the world's largest, constructed in aluminium, was a great achievement in structural engineering (Fig. 95). It provided a

Fig. 95 — Interior of the Dome of Discovery at the 1951 London Exhibition, showing aluminium alloy framework and roof covering.

unique example of the use of aluminium, tubular steel and concrete in building design. Even before this, the first aluminium bascule bridge in the world has been designed and built in Britain and had shown the value of prefabrication with ease of handling for such items.

To meet the requirements of torsional stiffness and resistance to local buckling of both beams and struts, without increasing their cross-sectional dimensions to the point of making the material uncompetitive with steel, ranges of sections reinforced with bulbs and fillets were developed by the Aluminium Development Association [5]. The basic research combined with experience gained permitted the compilation of a Code of Practice for the Structural Use of Aluminium in 1969 (with an updated version of 1987 — CP8118).

## 9.11 MINING ENGINEERING

The use of aluminium and its alloys in mines is subject to National Safety Regulations. Normally, aluminium and its alloys are found to be non-sparking and suitable for use with explosive material. However, investigations have shown that aluminium alloys having a magnesium content in excess of 1 per cent are liable to produce incendive sparks when brought into violent contact with rusty steel, creating an explosion hazard in a methane-bearing atmosphere.

Aluminium alloys with magnesium contents of less than 1 per cent have been used as castings for the body of lightweight manually operated tools and as extrusions for lightweight roof supports and as thick sheet or plate for the body of tubs for the transport of minerals and as plate and extrusions for hoists.

## 9.12 FOOD PROCESSING

The fact that aluminium is non-toxic and has high thermal conductivity suits it for use in the food processing industry as cooking utensils and other equipment (Fig. 96). It is also used extensively in domestic equipment, although in recent years it has been subject to competition from stainless steelware.

In the form of beverage cans, aluminium alloy 5182 finds an expanding market. This is an aluminium–magnesium alloy which combines high mechanical properties to allow a minimum can end thickness, a high degree of formability, and a minimum loss of strength on stoving after lacquering which is required to ensure a proper shelf life for beer or soft drinks [6].

## 9.13 DOMESTIC, OTHER THAN HOLLOWARE

Aluminium is used for refrigerator cooling panels and for the cabinets, for washing machines, parts of vacuum cleaning equipment, the framework of furniture and garden furniture, heat and light reflectors, anodised ornamental articles and sports equipment.

Aluminium containers for frozen foods, baking products, takeaway foods, etc., are a strong growth market. Foil is used for cheese wrapping and

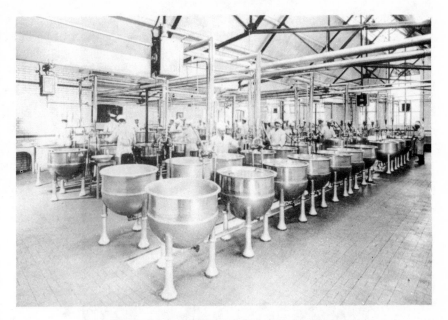

Fig. 96 — Aluminium food processing vessels.

foil laminate is used for a number of other foodstuffs; it is also supplied in the form of small coils a few metres in length for domestic use in culinary operations.

## 9.14  OTHER APPLICATIONS

Aluminium–magnesium alloys have been used for coinage and for nails for use with plaster board. One of the earliest uses of the metal was as a semi-precious metal and it is now used for costume jewellery capable of being anodised and coloured (Fig. 97).

Aluminium–magnesium alloy drawn wire has been used for chainlink fencing and as barbed wire, and in North America, wire is used for the weaving of fine mesh fly screens.

Aluminium–copper–magnesium–zirconium  and  aluminium–zinc–cal-cium are two sheet materials which have enhanced ductility at elevated temperatures but have metallic properties at room temperature. The elevated temperature properties make it possible to employ more complex 'one piece' shapes than with other production methods such as drop-hammer and rubber pressings [6].

Aluminium alloys are used for lightweight pressure vessels and gas cylinders.

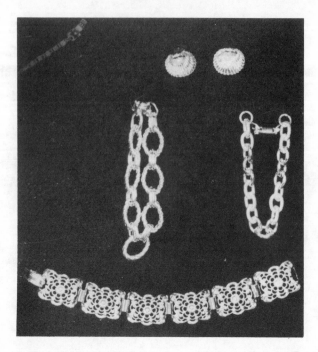

Fig. 97 — Examples of early aluminium costume jewellery.

In the chemicals' industry, uses include processing equipment, fatty acid condensers and shipping containers such as acetic acid drums.

## 9.15  ALUMINIUM AS AN ALLOYING ELEMENT

### 9.15.1  Copper alloys

*Aluminium bronzes* are a series of copper alloys having contents of up to 10.5 per cent aluminium which may be combined with additions of nickel, iron, manganese or silicon. As a group the aluminium bronzes have a high level of strength and toughness, good fatigue characteristics and reasonable properties at both high and low temperatures. Some of the alloys are amenable to heat treatment analogous in certain respects to steel.

The aluminium bronzes have the inherent chemical properties of copper enhanced by a tenacious surface film of oxides, mainly alumina but with some copper oxide. The oxide film is removed by alkalis and by strong acids in certain concentrations but is responsible for the aluminium bronzes being one of the most corrosion resistant copper-base materials [7]. The most important uses are propellers and stern-gear, valves (including valves in liquid oxygen installations), pumps, castings associated with heat exchangers, compressors, submarine periscopes, etc. Non-marine applications include papermaking machinery [8].

*Aluminium brass* (CZ110) shows better corrosion/erosion resistance in moving aerated sea water than Naval Brass (CZ112) or 70/30 Admiralty Brass (CZ111) — hence its use for condenser tubes.

### 9.15.2  Magnesium alloys
Aluminium is used as an alloying addition to magnesium in amounts of 3.5, 6.5, 8 or 9.5 per cent in combination with 0.5–1.15 per cent zinc and 0.37 per cent manganese.

## 9.16  ALUMINIUM ADDITIONS TO STEEL
Small amounts of aluminium in the form of coarse granules or small pieces added to molten steel act as a deoxidiser to yield a 'killed' steel.

## 9.17  ALUMINIUM POWDERS
There are two commercial methods for the production of aluminium powders:

(i) atomising — blowing a thin stream of molten metal through a small orifice by means of pressurised air. This produces fine metallic particles which are exhausted through ducting where they are cooled. These particles are separated from the air stream by centrifugal force by passage through a cyclone.
(ii) thin foil, which is subjected to attrition by stamping with steel balls in a ball mill.

The powders produced by these methods are graded by sieving. The finer grades are subjected to further treatment with suitable additives and solvents for applications as paints or printing inks.

The coarser grades are used for applications such as constituents as thermionic mixtures for welding or pyrotechnics or for explosives.

In recent years a number of new methods for the production of metal powders have been the subject of much research and development; these include RSP (rapid solidification process) [9] and RSPDP (rapid solidification plasma deposition process [10]. Table 56 gives a comparison of the properties of a conventional high strength alloy with laboratory RSP Al alloys.

Conventional aluminium alloys produced by ingot metallurgy lack appreciable thermal stability, losing strength rather rapidly between 100 and 200°C, and they have low elastic stiffness (modulus $70 \times 10^3$ MN/mm$^2$). The

**Table 56** [11] — Comparison of the properties of a conventional high strength aluminium alloy with laboratory RSP aluminium alloys

| Alloys | Ultimate tensile strength (MN/mm$^2$) | Modulus ($10^3 \times$ MN/mm$^2$) | Density (g/cm$^3$) |
|---|---|---|---|
| Conventional 7075 T-76 | 503 | 72 | 2.80 |
| RSP Al–Li-type alloy | 552 | 80 | 2.41 |
| RSP Al–Fe–Mn–Ni | 655 | 99 | 2.99 |

driving force behind the application of RSP to aluminium alloys is two-fold: to increase the operating temperature limit, and to increase strength and elastic modulus without degrading other properties [12]. Data reported on Al–8 per cent Fe/3.4 per cent Ce alloy, produced as microcrystalline powders with a high volume fraction of finely dispersed intermetallics, cold isostatically compacted, vacuum hot pressed and closed die forged to 15 mm diameter × 25 mm high (4.5 in × 1 in) pancakes, demonstrated encouraging capabilities with tensile strengths and stress rupture properties superior to the conventional A2219-T6 alloy at 343°C. The RSP Al–Fe–Ce alloys are regarded as strong candidates for impellers on auxiliary gas turbines [13]. The Al–Fe–Mo alloys produced from RSP powders have yield and ultimate strengths at 343°C comparable with conventional alloys at 177°C [11].

We are witnessing the beginning of a renaissance in alloy development after the comparative lull of the last twenty years. Undoubtedly further (RSP) alloys will emerge with unusual or unexpected combinations of properties.

## REFERENCES

[1] Bailey, J. C., *Metals and Materials*, Oct. 1981, p. 34.
[2] Brimelow, E. I., *Aluminium in Building*, p. 90, Macdonald, London.
[3] Brimelow, E. I., *Aluminium in Building*, p. 68.
[4] Hawkins, J. F., *The Metallurgist & Materials Technologist*, Aug. 1984, p. 412.
[5] Brimelow, E. I., *Aluminium in Building*, p. 169.
[6] Woodward, A. R., *The Metallurgist & Materials Technologist*, Jan. 1984, p. 23.
[7] West, E. G., *Copper and its Alloys*, 1982, Ellis Horwood Ltd, p. 110.
[8] Meigh, H. J., *The Metallurgist & Materials Technologist*, Oct. 1979, p. 583.

[9] Jones, H., *Rapid Solidification of Metals and Alloys*, The Institute of Metallurgists, 1982.
[10] Jackson, M. R., Rairdean, J. R., Smith, J. S. and Smith, R. W., *J. Met.*, 1981, **33**, (11), 23–27.
[11] Prince, A., *The Metallurgist & Materials Technologist*, Oct. 1984, p. 506.
[12] Millan, P. P., *J. Met.*, 1983, **35** (3), 76–81.
[13] Cox, A. R. and Bullman, L. S., ASME New York, Paper 1982, GT-77.

# Appendices

**1. RESISTANCE OF ALUMINIUM AND ITS ALLOYS TO ATTACK UNDER GIVEN CONDITIONS**

| Agent | Corrosive conditions | | Metal | | |
|---|---|---|---|---|---|
| | Temperature °C | Other conditions | Excellent | Satisfactory | Restricted applications |
| *Element* | | | | | |
| Arsenic | | Black | | Al 99.0–99.5% | |
| Bismuth | | Molten | Al 99.0–99.5% | Al 99.0–99.5% | |
| Boron | | Solid | Al 99.0–99.5% | | |
| Carbon | <650 | Dry solid | Al 99.0–99.5% | | |
| Iodine | 20 | Dry solid | Al 99.0–99.5% | | |
| | 20 | Solution | Al199.0–99.5% | Al alloys | |
| Lead | | Molten | Al 99.0–99.5% | | |
| Nitrogen | High | | Al 99.0–99.5% | | |
| Oxygen | <250 | | Al 99.0–99.5% | Al 99.0–99.5% | Al 99.0–99.5% |
| | 250–500 | | | | |
| Ozone | 20 | Dry | Al > 99.5 | — | |
| Phosphorus | | | — | — | Al > 99.5% |
| Selenium | | | — | — | Al 99.0–99.5% |
| Silicon | High | | — | Al 99.0–99.5% | |
| Sodium and potassium | | Dry solid | — | Al 99.0–99.5% | |
| Sulphur | | | Al 99.0–99.5% | Al 99.0–99.5% | |
| Tellurium | <600 | | Al 99.0–99.5% | Al 99.0–99.5% | |
| *Inorganic Compounds* | | | | | |
| Alum | 20 | 10% solution | — | Al > 99.5% | |
| | | | | Al 99.0–99.5% | |
| | | | | Al–Mg alloys | |
| | | | | Al–Mn alloys | |
| Aluminium chloride | 20 | Anhydrous | — | Al 99.0–99.5% | |
| Aluminium sulphate | | Solution | — | Al alloys | Al > 99.5% |
| Ammonia | | Moist gas | | Al–Mg alloys | |

| Reagent | Temperature (°C) | State | Alloys |
|---|---|---|---|
| Ammonium chloride | 20–70 | Solution | Al > 99.5%; Al–Mn alloys |
| Ammonium nitrate | 20 | Solution | Al–Mg alloys |
| Ammonium phosphate | 20 | Solution | Al–Mg alloys |
| Ammonium sulphate | 20 | Solution | Al > 99.5%; Al–Mn alloys |
| Ammonium sulphide | 20–100 | Solution | Al 99.0–99.5%; Al–Mg alloys; Al > 99.5% |
| Antimony trichloride | 0–100 | Solution | Al > 99.5% |
| Arsenious oxide | — | Solution | Al > 99.5% |
| Barium hydroxide | — | Solution | Al 99.0–99.5% |
| Boric acid | — | Solution | Al 99.0–99.5%; Al–Mn alloys |
| Calcium carbide | — | Dry | Al 99.0–99.5% |
| Calcium chloride | 20–100 | Solution | Al > 99.5%; Al–Mn alloys |
| Calcium hydroxide | 20 | Solution | Al > 99.5%; Al–Mg alloys |
| Calcium nitrate | — | Solution | Al > 99.5% |
| Calcium sulphate | 20 | Solution | Al > 99.5%; Al–Mg alloys |
| Carbon bisulphate | 20–b.p. | Liquid | Al–Mn alloys |
| Carbon dioxide | 20 | Solution | Al 99.0–99.5%; Al > 99.5% |
| Carbon monoxide | <500 | Gas | Al–Mn alloys; Al 99.0–99.5%; Al alloys |
| Chromic acid | 20 | Solution | Al > 99.5% |
| Chrome alum | — | Solution | Al–Mn alloys |

| Agent | Corrosive conditions | | Metal | | |
|---|---|---|---|---|---|
| | Temperature °C | Other conditions | Excellent | Satisfactory | Restricted applications |
| Ferrous sulphate | 20 | Solution | | Al > 99.5% Al–Mn alloys | |
| Hydrogen peroxide | — | Pure Solution | | Al–Mn alloys Al 99.0–99.5% Al–Mg alloys | |
| Hydrogen sulphide | Cold Hot | Dry gas Dry gas | { Al–Mn alloys Al 99.0–99.5% Al–Mg alloys | | |
| | Cold Cold Hot | Moist gas Solution Solution | | { Al–Mn alloys Al 99.0–99.5% Al–Mg alloys | |
| Lead oxide | 20 | Solid | Al > 99.5% | | |
| Magnesium carbonate | 20 | Sat. solution and paste | | Al–Mg alloys Al 99.0–99.5% | |
| Magnesium chloride | 20 | Solution | | Al > 99.5% | |
| Magnesium sulphate | 20 | Solution | | Al > 99.5% | |
| Nitric acid | | <80% solution | | Al > 99.5% Al–Mn alloys | |
| | | >80% solution | | Al > 99.5% Al–Mn alloys | |
| | | Fuming | | Al–Si alloys Al 99.0–99.5% Al–Mg$_2$Si alloys | |
| Nitrogen oxides Nitrous acid | | Dry gas Solution | Al > 99.5% | Al > 99.5% Al–Mn alloys | |

| | | | Al–Mn alloys | |
|---|---|---|---|---|
| Potassium bromide | | Solution | | Al > 99.5% / Al–Mg alloys / Al 99.0–99.5% |
| Potassium carbonate | | Solution | | |
| Potassium chlorate | | Solution | | Al–Mn alloys / Al99.0–99.5% |
| Potassium ferrocyamide | | Solution | | Al–Mn alloys / Al > 99.5% |
| Potassium iodide | 20–100 | Solution | | |
| Potassium nitrate | | Solution | Al > 99.5% / Al–Mn alloys / Al > 99.5% | Al alloys / Al 99.0–99.5% / Al–Mg alloys |
| Potassium sulphate | 20 | Molten | | |
| | | Solution | Al > 99.5% | |
| Potassium thiocyanate | 20–100 | Solution | Al–Mg alloys / Al 99.0–99.5% | Al–Mn alloys / Al 99.0–99.5% / Al > 99.5% |
| Sodium bicarbonate | 20 | Solution | | Al–Mn alloys / Al > 99.5% |
| Sodium bisulphate | | | | Al–Mn alloys / Al > 99.5% |
| Sodium borate | | Solution | | Al–Mn alloys |
| Sodium dihydrogen phosphate | | Solution | Al 99.0–99.5% | Al–Mn alloys |
| Sodium hydrogen phosphate | | Solution | | Al–Mn alloys |
| Sodium meta phosphate | | | | |
| Sodium silicate | | | | Al–Si alloys / Al 99.0–99.5% |
| Sodium sulphite | 20 | Solution | Al–Mn alloys / Al 99.0–99.5% | |
| Sodium thiosulphate | 20 | Solution | Al 99.0–99.5% | |

| Agent | Corrosive conditions Temperature °C | Other conditions | Metal Excellent | Satisfactory | Restricted applications |
|---|---|---|---|---|---|
| Sulphur dioxide | | Dry gas | Al-Mn alloys<br>Al 99.0–99.5% | | |
| Sulphuric acid | 20 | Solution | | Al > 99.5% | Al > 99.5% |
| Sulphurous acid | 20 | Pure solution | | Al-Mn alloys | |
| Sulphuryl chloride | b.p. | Pure | | Al > 99.5% | Al > 99.5% |
| Zinc chloride | | | | | |
| Zinc sulphate | 20 | Solution | | Al > 99.5%<br>Al-Mn alloys | |
| *Organic Compounds* | | | | | |
| Acetanilide | | | Al-Mn alloys<br>Al 99.0–99.5% | | |
| Acetic acid | 0–118 | 0–100%<br>Solution | | | Al > 99.5%<br>Al-Mn alloys |
| Acetic anhydride | | Pure | | Al-Mn alloys<br>Al 99.0–99.5%<br>Al-Si alloys | |
| Acetylene | | Gas | | Al 99.0–99.5%<br>Al alloys | |
| Alcohol (ethyl) | 20 | Pure | | Al-Mn alloys<br>Al 99.0–99.5% | |
| Aldehydes (aliphatic) | 20 | Solution | | Al > 99.5%<br>Al-Mn alloys<br>Al > 99.5%<br>Al-Mn alloys | |

| Substance | Temp (°C) | Condition | Aluminium grade |
|---|---|---|---|
| Alakloids | | | |
| Aluminium acetate | 20 | Solution | Al > 99.5% |
| Amyl acetate | 20 | Anhydrous | Al > 99.5% |
| Amyl alcohol | | | Al–Mn alloys; Al 99.0–99.5% |
| Aniline | 20 | Pure | Al–Mn alloys; Al–Si alloys; Al 99.0–99.5% |
| Benzene | | | Al–Mn alloys; Al 99.0–99.5% |
| Benzoic acid | | Solid | Al > 99.5% |
| Benzoic acid | | Solution | Al > 99.5%; Al–Mn alloys |
| Butyl acetate | | | Al–Mn alloys; Al 99.0–99.5% |
| Butyl alcohol | 20–b.p. | Liquid | Al–Mn alloys; Al 99.0–99.5% |
| Butyric acid | | 0–100% | Al > 99.5% |
| Butyric acid | | Solution | Al–Mn alloys; Al–Si alloys |
| Camphor | 20–b.p. | | Al–Mn alloys; Al 99.0–99.5% |
| Carbon tetrachloride | 20 | Wet | Al 99.0–99.5%; Al alloys |
| Citric acid | 20–100 | Solution | Al–Mn alloys; Al > 99.5% |
| Ethyl acetate | | | Al–Mn alloys |
| Fatty acids (higher) | 20–b.p. | Commerical | Al > 99.5%; Al–Mn alloys |
| Formaldehyde | 20–100 | Solution | Al > 99.5% |
| Formic acid | | Solution | Al 99.5%; Al–Mn alloys |
| Furfurol | | | Al–Mn alloys; — |

| Agent | Corrosive conditions | | Metal | | |
| --- | --- | --- | --- | --- | --- |
| | Temperature °C | Other conditions | Excellent | Satisfactory | Restricted applications |
| Gallic acid | | Solution | | Al–Mn alloys / Al 99.0–99.5% | |
| Hydrazine | 20 | Anhydrous | | Al > 99.5% | |
| Indole | | | | Al–Mn alloys / Al 99.0–99.5% | |
| Iodoform | | Vapour | Al–Mn alloys / Al 99.0–99.5% | | |
| Isatin | | | | Al–Mn alloys / Al 99.0–99.5% | |
| Ketones (aliphatic) | | | Al > 99.5% / Al–Mn alloys / Al–Mg alloys | | |
| Lactic acid | 20 | Solution | | Al > 99.5% | |
| Malic acid | | | | Al–Mn alloys | |
| Methyl alcohol | | Solution | | Al > 99.5% | |
| Methyl chloride | | | | | Al > 99.5% |
| Naphthols | | | | Al > 99.5% / Al–Mn alloys | |
| Naphthylamine | | | | Al > 99.5% | |
| Nitrobenzene | 20 | | | Al–Mn alloys / Al 99.0–99.5% | |
| Oxalic acid | 20 | Solution | | Al > 99.5% / Al–Mn alloys | |
| Paraldehyde | 20 | | Al–Mn alloys / Al 99.0–99.5% | | |
| Phenol | | | | | Al–Mn alloys / Al 99.0–99.5% |

| Substance | Temp. | Form | Alloy |
|---|---|---|---|
| Potassium oxalate | 20 | Molten | Al 99.0–99.5% Al > 99.5% |
| Quinone | | Solution | Al 99.0–99.5% |
| Salicylic acid | 20–80 | Solution | Al > 99.5% |
| Sodium tartrate | 20 | Solution | Al > 99.5% |
| Starch | | | Al > 99.5% |
| | | | Al–Mn alloys |
| | | | Al 99.0–99.5% |
| Succinic acid | 20–100 | Solution | Al > 99.5% |
| | | | Al–Mn alloys |
| Tannic acid | | | Al–Mn alloys |
| | | | Al 99.0–99.5% |
| Tartaric acid | | Solution | Al > 99.5% |
| Thrichloracetic acid | | | |
| Trichlorethylene | 20 | Liquid | Al > 99.5% Al 99.0–99.5% Al alloys |
| *Miscellanous* | | | |
| Asphalt | | | Al99.0–99.5% |
| Atmosphere rural | 20 | | Al > 99.5% |
| | | | Al–Mn alloys |
| | | | Al–Cu alloys |
| | | | clad with Al > 99.5% |
| | | | Al–Mg alloys |
| | | | AlMg$_2$Si alloys |
| | | | Al > 99.5% |
| | | | Al–Cu alloys |
| | | | clad with |
| | | | Al > 99.5% |
| | | | Al–Mg alloys |
| | | | AlMg$_2$Si alloys |
| Atmosphere (urban or industrial) | | | Ai 99.0–99.5% |

| Agent | Corrosive conditions | | Metal | | |
| --- | --- | --- | --- | --- | --- |
| | Temperature °C | Other conditions | Excellent | Satisfactory | Restricted applications |
| Atmosphere (marine) | | | | Al–Si alloys<br>Al–Cu alloys<br>Al > 99.5%<br>Al–Cu alloys clad with Al > 99.5%<br>Al–Mg alloys<br>Al–Mn alloys<br>Al 99.0–99.5%<br>AlMg$_2$Si alloys<br>Al–Cu alloys | |
| Dye stuff — vat (acid) (basic) | | Solution | | | |
| Fruit juices | | Natural and some with preservatives | Al > 99.5% | Al > 99.5% | Al > 99.5%<br>Al–Mn alloys |
| Lacquers<br>Linseed oil | b.p. | | Al–Mg alloys<br>Al > 99.5% | Al > 99.5% | |
| Margarine<br>Milk<br>Paraffin | | | Al–Mn alloys<br>Al 99.0–99.5% | Al > 99.5%<br>Al 99.0–99.5% | Al–Mn alloys |
| Petroleum | | | | Al–Mn alloys<br>Al 99.0–99.5% | |
| Petroleum oils crude | <250<br>250–500 | | | Al 99.0–99.5%<br>Al–Mn | |

| Material | Alloys |
|---|---|
| Resins | |
| Sea water | Al 99.0–99.5%<br>Al > 99.5%<br>Al–Mg alloys<br>Al–Mn alloys<br>Al 99.0–99.5%<br>clad Al–Cu alloys<br>Al–Si alloys<br>Al–Mg$_2$ Si alloys |
| Sewage | Al–Mn alloys<br>Al 99.0–99.5%<br>Al–Mg$_2$ Si alloys |
| Soap | Al–Mg alloys<br>Al 99.0–99.5%<br>Al > 99.5% |
| Sugar | |
| Tanning (solutions) | 20 |
| Water (fresh tap) | Al > 99.5% |
| Wine | Al > 99.5% |

Appendix based on material from several sources.

## 2.  MACROGRAPHIC AND MICROGRAPHIC EXAMINATION

### Macrographic examination

This is used to check whether the cast or recrystallised material has the desired fine grain structure and to reveal the grain flow in extruded and forged components. Surface flaws, porosity, shrinkage cavities, local segregations and non-metallic inclusions may also be determined by macrographic examination.

For macrographic examination of cross-sections of castings, forgings and extrusions, the face of the specimen to be examined should be machined to a fine finish by turning or milling.

A suitable etchant is [1]:

>     20 ml conc hydrochloric acid     38%
>     20 ml conc nitric acid           65%
>      5 ml conc hydrofluoric acid     30%
>     60 ml water

For those alloys (Al–Zn, Al–Cu) which are attacked too strongly by this etchant, it is diluted with more water.

The examination of specimens of extruded products for defects may be conducted by immersion in a 10 per cent solution of NaOH at 70°C, washing in water, followed by a dip in 20 per cent by vol $NHO_3$ washing in water and drying.

To reveal surface flaws in sheet metal anodising in sulphuric acid or etching in sulphuric acid, 10 per cent heated to 50°C may be employed.

Foil is best etched with 100 cc 5 per cent aqueous ferric chloride solution and 5–8 drops concentrated hydrofluoric acid.

### Micrographic examination

The preparation of specimens for micrographic examination requires great care. Small specimens of sheet and wire are embedded in a thermosetting resin for machining, polishing and etching.

Suitable etchants are:

(a)  to reveal the alloying constituents:
>     10 drops 10 per cent hydrofluoric acid
>     10 drops conc nitric acid
>     100 ml distilled water
>     Duration of etching: 15–60 s;

(b)  to etch the grain boundaries:
>     10 drops 30 per cent hydrofluoric acid
>     10 drops conc nitric acid
>     3 ml 10 per cent ferric chloride solution
>     100 ml distilled water
>     Duration of etching 1–4 mins

## 3. ALUMINIUM COMPOUNDS

In addition to the uses of aluminium already discussed, many compounds of the metal are important in a number of industries and a brief summary is given as a guide to further study.

The metal itself is used in a number of industries because of its chemical properties: for example, for the deoxidation of steel melts and as an addition to other metals as an alloying element, especially the aluminium–bronzes which are copper-base alloys containing up to about 10 per cent aluminium.

In combination with ammonia it forms a powerful explosive known as Ammonal. In the 'Thermit' process, finely divided aluminium is mixed with iron oxide and when ignited the exothermic reaction produces molten iron and aluminium oxide as a slag. The process is thus used for welding steel components, such as rails for train and tram tracks.

Alumina — the peroxide of aluminium, $Al_2O_3$ — is an important commercial material used in a number of forms in many industries. It is also sold as hydrate of aluminium which is the base for the manufacture of aluminium sulphate $(Al_2(SO_4)_318H_2O)$ required by the paper industry and for water purification including the treatment of sewage and trade effluents. It is used too in the manufacture of catalysts for petroleum refining. It is effective as a means of increasing the acidity of soil and to prevent dust from concrete floors. Smaller quantities are used for colours and as a mordant in dyeing where it combines particularly with vegetable colours to form insoluble compounds known as 'lakes', thus fixing the colours to make them fast in washable fabrics, fabric printing and water proofing as well as in tanning.

Aluminium hydroxide is a constituent of a number of pharmaceutical products because of its ant-acid properties which are particularly useful, as the oxide is amphoteric. The oxide has extensive applications as a constituent of glasses, and calcined alumina is applied in place of silica for the bonding of pottery during firing.

Fused alumina has increased in importance in recent years as the major constituent of some high duty refractories and ceramics of increasing interest for new industrial applications. It is a heat resistant electrical insulator: for example, its extensive use for sparking-plug bodies. It is a heat resisting constituent of some parts of arc welding torches, and because it is relatively inert it is applied in various types of chemical plant.

Many naturally occurring forms of alumina are well known as gem stones, in particular rubies, sapphires, emeralds, topaz and diamantine (which is also used as a fine abrasive). With aluminium phosphate it is known as turquoise and most of these gem stones are now made artificially by electric furnace techniques.

Another natural form of alumina is corundum which is available in the earth's crust as high density crystals (specific gravity around 4): it is also made synthetically. 'Alundum' and 'Aloxite' are forms of hard aluminium oxide made by fusing bauxite in electric furnaces. The hardness of these types of alumina is around 9 on the Moh scale and they are thus used extensively as abrasives, for example, on cloth or papers for finishing and polishing.

Emery is a natural form of hard alumina coated with oxides of iron and manganese and is also used widely as an abrasive material for grinding wheels and polishing elements.

Activated alumina is a particularly dehydrated aluminium hydroxide with a permanent physical structure of high porosity. It is thus useful for the removal of water vapour from mixtures of gases and also for the successive absorption of certain gases particularly of substances of high boiling point. It is also used as a catalyst.

The single sulphate of aluminium ($Al_2(SO_4)_318H_2O$) is manufactured in large quantities by the action of sulphuric acid on bauxite followed by purification, for use in the textile, water treatment, paper making, leather and soil treatment industries. The double sulphates form a separate group.

Alums are the double sulphates of the alkali metals with either aluminum, iron or chromium, with today's general formula $M^+Al^{3+}(SO_4^{2-})_2.12H_2O$ which differs from the former monobasic sulphate: tribasic sulphate: $24H_2O$. They are water soluble and are extensively used industrially.

They have been known since Pliny named the double aluminium sulphate as 'alum' from which was derived the oxide, alumina. Indeed from ancient times, the best known member of the group has been Roman Alum or the double sulphate of aluminium and potassium ($KAl(SO_4)_212H_2O$ or $K_2SO_4Al_2(SO_4)_3.24H_2O$).

Soda alum, $NaAl(SO_4)_212H_2O$ is a useful mordant in dyeing since it forms aluminium hydroxide by hydrolysis, whilst chrome alum with no aluminium content, $KCr_2(SO_4)_212H_2O$, is valuable as a tanning agent in making chrome leather. Ammonium alum, $NH_4Al(SO_4)_2$, is also made on a large scale.

Other salts used in textile printing and dyeing are aluminium acetate — $Al_2(C_2H_3O_2)_3$ — ('red liquor'), aluminium fluoride — $Al_2F_67H_2O$ — applied with small amounts of sulphate as a moth-proofing compound especially for wool. Aluminium borate ($2Al_2O_3H_2O$) and the insoluble phosphate ($AlPO_4$) are used in ceramics. Other compounds used industrially include aluminium resinate in the sizing of paper and aluminium stearate in the oil and paint trades for thickening and increasing the viscosity of oils in which it is dissolved.

Aluminium nitrate is used in the leather and textile industries. Aluminium chloride is a volatile solid employed as a catalyst in petroleum cracking plants.

## 4. SI UNITS USED IN THE INDUSTRIAL METALS SERIES

| Quantity | SI unit | Recommended mutiples | Other related units (or names) |
|---|---|---|---|
| Length | m (metre) | km<br>mm<br>m<br>nm | $\text{Å} = 10^{-10}\,\text{m}$<br>$= 10^{-7}\,\text{mm}$ |
| Area | $m^2$ | $mm^2$ | hectare (ha) $= 10^4\,m^2$ |
| Volume | $m^3$ | $mm^3$ | litre (l) |
| Time | s (second) | ms<br>s<br>ns | minute (min)<br>hour (h)<br>day (d) |
| Mass | kg | g<br>mg<br>$\mu$g | tonne (t) $= 10^3\,\text{kg}$<br><br>metric carat $= 2 \times 10^{-4}$ kg |
| Density | $\dfrac{kg}{m^3}$ | $\dfrac{g}{cm^3}$ | |
| Force | N (Newton) | MN<br>kN | dyne/cm $= 10^{-5}$ N/cm |
| Impact strength | $\dfrac{J}{m^2}$ | $\dfrac{kJ}{m^2}$ | $\dfrac{J}{cm^2}$ |
| Stress and pressure | $\dfrac{N}{m^2}$ | $\dfrac{MN}{m^2}$<br><br>$\dfrac{N}{mm^2}$ | hectobar (hbar)<br>$= 10^7/Nm^2$<br><br>Pascal $= N/m^2$ |

| Quantity | SI unit | Recom-mended mutiples | Other related units (or names) |
|---|---|---|---|
| Viscosity | $\dfrac{NS}{m^2}$ | | centipoise (cP) $= 10^{-3}NS/m^2$ |
| Surface tension | $\dfrac{N}{m}$ | $\dfrac{mN}{m}$ | |
| Energy and work | J (joule) | GJ MJ kJ | kilowatt hour (kWh) $= 3.6 \times 10^6$ J $= 3.6$ MJ electronvolt (eV) $(1.60210 \pm 0.00001 \times 10^{-19}$ J) |
| Power | W (watt) | GW MW kW | $\dfrac{IJ}{s}$ |
| Temperature | K | | °C (Celsius) |
| Linear expansion coefficient | $\dfrac{1}{K}$ | | $\dfrac{1}{°C}$ |
| Heat quantity | J | MJ kJ mJ | |
| Thermal conductivity | $\dfrac{W}{mK}$ | | $\dfrac{W}{m°C}$ |
| Specific heat | $\dfrac{J}{kgK}$ | $\dfrac{kJ}{kgK}$ | $\dfrac{J}{kg°C}$ |
| Latent heat | $\dfrac{J}{kg}$ | $\dfrac{kJ}{kg}$ | |
| Electric current | A (ampere) | mA | |

| Quantity | SI unit | Recommended mutiples | Other related units (or names) |
|---|---|---|---|
| Electric potential | V (volt) | MV<br>kV<br>mV | |
| Current density | $\dfrac{A}{m^2}$ | $\dfrac{A}{cm^2}$<br>$\dfrac{A}{mm^2}$ | |
| Resistance | Ω (ohm) | mΩ<br>μΩ | |
| Resistivity | Ωm | μΩm | $\mu\Omega cm = 10^{-8}\ \Omega m$ |

## 5. COMMON CONVERSION FACTORS

| | |
|---|---|
| 1 yard | = 0.9144 m |
| 1 foot | = 0.3048 m |
| 1 inch | = 25.4 mm |
| 1 mil | = 0.001 in = 25.4 $\mu$m |
| 1 yd$^2$ | = 0.83617 m$^2$ |
| 1 ft$^2$ | = 0.092903 m$^2$ |
| 1 in$^2$ | = 6.4516 cm$^2$ |
| 1 yd$^3$ | = 0.764555 m$^3$ |
| 1 ft$^3$ | = 28.3168 dm$^3$ |
| 1 in$^3$ | = 16.3871 cm$^3$ |
| 1 gal (Imp.) | = 4.5461 litre or |
| | = 4.54609 dm$^3$ |
| 1 in$^4$ | = 41.6231 cm$^4$ |
| (Moment of Section) | |
| 1 troy oz | = 31.1035 g |
| 1 dwt | = 1.55517 g |
| (pennyweight) | |
| 1 oz (av) | = 28.3495 g |
| 1 lb | = 0.453592 kg |
| 1 cwt | = 50.802 kg |
| 1 ton | = 1016.05 kg |

| 1 tonne (metric) | = 1000 kg |
|---|---|
| 1 lb/in$^3$ | = 27.6999 g/cm$^3$ |
| 1 lbf | = 4.448.22 N |
| 1 tonf | = 9964.02 N |
| 1 tonf/in$^2$ | = 15.4443 MN/m$^2$ or = 15.4443 N/mm$^2$ |
| 1 assay ton | = 32.6667 g |

**Additional conversion factors**

| | *C.G.S units* | *SI units* |
|---|---|---|
| Wavelength | 1 Å | $= 10^{-1}$ nm |
| Heat energy | 1 cal | = 4.187 J |
| Heat capacity | 1 cal/°C g mole | = 4.187 J/K mol |
| Heat content | 1 cal/g mole | = 4.187 J/mol |
| Thermal conductivity | 1 cal/cm S °C | = 418.7 W/mk |
| Pressure | 1 mm Hg | = 133.3 Pa |
| Pressure | 1 torr | = 133.3 Pa |
| Pressure | 1 bar | $= 10^5$ Pa |
| Pressure | 1 atmosphere | $= 1.013 \times 10^5$ Pa |
| Stress | 1 kgf/mm$^2$ | = 9.807 MN/m$^2$ |
| Electrical resistivity | 1 microhm cm | $= 10^{-8}$ m |
| Magnetic field | 1 oersted | = 79.58 A/m |
| Magnetic flux density | 1 gauss | $= 10^{-4}$ T |
| Magnetic permeability | 1 gauss/oersted | $= 12.57 \times 10^{-7}$ H/m |

# Index